THE
STORY
OF
SCIENCE

THE
STORY
OF
SCIENCE

From the WRITINGS *of* ARISTOTLE
to the BIG BANG THEORY

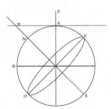

SUSAN WISE BAUER

W. W. NORTON & COMPANY

NEW YORK LONDON

For information about permission to reproduce selections from this book,
write to Permissions, W. W. Norton & Company, Inc.,
500 Fifth Avenue, New York, NY 10110

For information about special discounts for bulk purchases, please contact
W. W. Norton Special Sales at specialsales@wwnorton.com or 800-233-4830

Manufacturing by RRD Harrisonburg
Book design by JAM Design
Production manager: Julia Druskin

Library of Congress Cataloging-in-Publication Data
Bauer, S. Wise.
The story of science ; from the writings of Aristotle to the big bang theory / Susan
Wise Bauer. — First edition.
pages cm
Includes bibliographical references and index.
ISBN 978-0-393-24326-0 (hardcover)
1. Science—History. I. Title.
Q125.B3285 2015
509—dc23
2015000136

W. W. Norton & Company, Inc.
500 Fifth Avenue, New York, N.Y. 10110
www.wwnorton.com

W. W. Norton & Company Ltd.
Castle House, 75/76 Wells Street, London W1T 3QT

1 2 3 4 5 6 7 8 9 0

The Books

The *Aphorisms* of Hippocrates (ca. 420 BC)
Plato, *Timaeus* (ca. 360 BC)
Aristotle, *Physics* (ca. 330 BC)
Aristotle, *History of Animals* (ca. 330 BC)
Archimedes, "The Sand-Reckoner" (ca. 250 BC)
Lucretius, *On the Nature of Things* (ca. 60 BC)
Ptolemy, *Almagest* (ca. AD 150)
Nicolaus Copernicus, *Commentariolus* (1514)

Francis Bacon, *Novum organum* (1620)
William Harvey, *De motu cordis* (1628)
Galileo Galilei, *Dialogue concerning the Two Chief World Systems*
   (1632)
Robert Boyle, *The Sceptical Chymist* (1661)
Robert Hooke, *Micrographia* (1665)
Isaac Newton, *Philosophiae naturalis principia mathematica*
   (1687/1713/1726)

Georges-Louis Leclerc, Comte de Buffon, *Natural History:*
   General and Particular (1749–88)
James Hutton, *Theory of the Earth* (1785)
Georges Cuvier, "Preliminary Discourse" (1812)
Charles Lyell, *Principles of Geology* (1830)
Arthur Holmes, *The Age of the Earth* (1913)

Alfred Wegener, *The Origin of Continents and Oceans* (1915)
Walter Alvarez, *T. rex and the Crater of Doom* (1997)

Jean–Baptiste Lamarck, *Zoological Philosophy* (1809)
Charles Darwin, *On the Origin of Species* (1859)
Gregor Mendel, *Experiments in Plant Hybridisation* (1865)
Julian Huxley, *Evolution: The Modern Synthesis* (1942)
James D. Watson, *The Double Helix* (1968)
Richard Dawkins, *The Selfish Gene* (1976)
E. O. Wilson, *On Human Nature* (1978)
Stephen Jay Gould, *The Mismeasure of Man* (1981)

Albert Einstein, *Relativity: The Special and General Theory* (1916)
Max Planck, "The Origin and Development of the Quantum
    Theory" (1922)
Erwin Schrödinger, *What Is Life?* (1944)
[Edwin Hubble, *The Realm of the Nebulae* (1937)]
Fred Hoyle, *The Nature of the Universe* (1950)
Steven Weinberg, *The First Three Minutes: A Modern View of the
    Origin of the Universe* (1977)
James Gleick, *Chaos* (1987)

Contents

PART IV: READING LIFE
(With Special Reference to Us)

List of Illustrations

Acknowledgments

Thanks to Julia Kaziewicz for clearing permissions with her usual cheerful energy, and to Richie Gunn for bringing his own particular style to the illustrations.

To Greg Smith and Justin Moore for reading drafts and making meticulous, thoughtful suggestions (many of which I ignored, so any errors that remain are mine alone).

To Michael Carlisle at Inkwell and his very efficient assistant Hannah Schwartz for their expert management.

To the friends who have patiently listened to me talk about my most recent enthusiasm, for much longer than they probably wanted: Liz Barnes, Mel Moore, and Boris Fishman. Hope the wine helped.

Finally, as always, to Starling Lawrence at W. W. Norton, whose faint praise is worth more than most people's adulation. Also to the amazingly capable Ryan Harrington, and the rest of the Norton crew who have been helping to bring my projects to the finish line since 1999—including but certainly not limited to Francine Kass, Michael Levantino, Stephen King, Don Rifkin, Nancy Palmquist, Eugenia Pakalik, Golda Rademacher, Elizabeth Riley, Nomi Victor, and Joe Lops. And to Tracy Vega, Meg Sherman, Kristin Keith, and the rest of the wonderful Norton sales force: Thanks for all you do.

Preface

Or, How to use this book

> There is no human knowledge which cannot lose its
> scientific characterwhen men forget the conditions under
> which it originated,the questions which it answered, and the
> functions it was created to serve.
>
> —Benjamin Farrington,
> *Greek Science: Its Meaning for Us*

This is not a history of science.

Histories of science have been written, in great numbers (and at great length), by many other writers. They abound: studies of Greek science, Renaissance science, Enlightenment science, Victorian science, modern science, science and society, science and philosophy, science and religion, science and "the people."

Of course these histories have value. But somehow, the nature of science itself seems to get lost in the details. Most "people," regular citizens who have no professional training in the sciences, still have no clear view of what science does—or what it means.

Most of us are fed science in news reports, interactive graphs, and sound bites. These may give us a fuzzy and incomplete glimpse of the facts involved, but the ongoing science battles of the twenty-first century show that the facts aren't enough. Decisions that affect stem cell research, global warming, the teaching of evolution in elementary schools—these are being made by voters (or, independently, by their theoretical representatives) who don't actually understand why biologists think stem cells are important, or how environmental scientists came to the conclusion that the earth is warming, or what the Big Bang actually is (neither big nor a bang; see Chapter 27).

So this is a slightly different kind of history. It traces the development of great science *writing*—the essays and books that have most directly affected and changed the course of scientific investigation. It is intended for the interested and intelligent nonspecialist. It shows science to be a very human pursuit: not an infallible guide to truth, but a deeply personal, sometimes flawed, often misleading, frequently brilliant way of understanding the world.

Each part presents a chronological series of "great books" of science, from the most ancient works of Hippocrates, Aristotle, and Plato, all the way up to the modern works of Richard Dawkins, Stephen Jay Gould, James Gleick, and Walter Alvarez. The chapters provide all of the historical, biographical, and technical information you need to understand the books themselves, and conclude with recommended editions. Older works, which don't necessarily need to be read in their entirety, are also excerpted on this book's website; links are provided in the text. (The website also lists available e-book versions, many of which can be difficult to find for pre-twentieth-century volumes.)

This is by no means meant to be a comprehensive list of important books in science, and readers may quibble with my selections. Many worthy books in science are not on my lists (search for any "great books of science" list and you'll find hundreds). I chose these books not merely to highlight particular *discoveries* in science as such, but to illuminate the way we *think* about science. It is an interpretive list, not an exhaustive one.

Part I, "The Beginnings," covers the origins of science itself. Part II, "The Birth of the Method," explains how and why the scientific method that we now take for granted arose. The rest of the book is an introduction to major works in three different areas: the science of the earth, the science of life, and the science of the cosmos. The order is deliberate. Geology steered us toward the time frame that modern biology demands, and that time frame then led us to a new contemplation of the entire cosmos.

In Parts III–V, alert readers will notice a shift: sometime after the 1940s, the "classics" listed are most often the books that made new theories or discoveries visible to the world, not the journal articles or conference papers that first introduced them to other scientists. So, to understand catastrophism, you will read Walter Alvarez's 1997 book *T. rex and the Crater of Doom* rather than the 1980 article "Extraterrestrial Cause for the Cretaceous-Tertiary Extinction" written by Alvarez and three coauthors; to understand the Big Bang, Steven Weinberg's best-selling *The First Three Minutes* rather than any of the (multiple) scientific papers about cosmic background radiation that preceded it.

After World War II, the practice of science became increasingly specialized.* Scientists gained academic recognition, the interest of their colleagues, and (occasionally) financial reward through the careful investigation of individual puzzle pieces, not through attempts to sketch entire scientific landscapes. Scientific theories were formed, evaluated, and supported or rejected by a scientific community that talked, more and more, to itself—and often in a language incomprehensible to outsiders. *The Double Helix* and *The Selfish Gene* are "great books" of biology in quite a different sense than is William Harvey's *De motu cordis*; Harvey could lay out his discoveries to his colleagues and the general public simultaneously, but neither James Watson nor Richard Dawkins could count on anyone outside an academic department to read his original paper. ("Parasites, Desiderata Lists and the Paradox of the Organism" reached a relatively small audience.) Instead, they had to popularize: synthesize, simplify, and explain.

Yet *The Double Helix*, *The Selfish Gene*, and *De motu cordis* all performed the same task: they opened up, for all of us, a new way of thinking about the natural world.

---

* This specialization had multiple causes; massive investment by Western industrialists in research projects that might yield commercial gains and the growing role of the university in nurturing (and paying) scientists are probably central, but other factors played a part as well. The phenomenon is beyond the scope of this book, but interested readers might want to consult John J. Beer and W. David Lewis, "Aspects of the Professionalization of Science," *Daedalus* 92, no. 4 (Fall 1963): 764–84; or Chapter 8 of I. Bernard Cohen, *Revolution in Science* (Harvard University Press, 1985).

·

You do not actually have to read every text I discuss. Pick the great books you want to start with. If you're most interested in biology, or cosmology, you don't have to read all of my recommended texts from Parts I and II before you jump into the recommended texts in Part IV or Part V.

But at the very least, read my chapters *about* the books and the ideas behind them. Scientists who grapple with biological origins are still affected by Platonic idealism today; Charles Lyell's nineteenth-century geological theories still influence our understanding of human evolution; quantum theory is still wrestling with Francis Bacon's methods.

To interpret science, we have to know something about its past. We have to continually ask not just "What have we discovered?" but also *"Why did we look for it?"* In no other way can we begin to grasp why we prize, or disregard, scientific knowledge in the way we do; or be able to distinguish between the promises that science can fulfill and those we should receive with some careful skepticism.

Only then will we begin to understand science.

·

A note on vocabulary: Throughout, I tend to use the terms "theory" and "hypothesis" interchangeably. A twenty-first-century scientist might point out that a theory is more comprehensive than a hypothesis, or longer-lived, or has stronger mathematical underpinnings. But both words refer to a theoretical structure that makes sense of evidence. Since it isn't always clear when a hypothesis becomes a theory, and since scientists in different centuries and different fields tend to use the words in different contexts, I have declined to struggle over the distinction.

PART

I

THE
BEGINNINGS

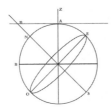

The *Aphorisms* of Hippocrates (ca. 420 BC)

Plato, *Timaeus* (ca. 360 BC)

Aristotle, *Physics* (ca. 330 BC)

Aristotle, *History of Animals* (ca. 330 BC)

Archimedes, "The Sand-Reckoner" (ca. 250 BC)

Lucretius, *On the Nature of Things* (ca. 60 BC)

Ptolemy, *Almagest* (ca. AD 150)

Nicolaus Copernicus, *Commentariolus* (1514)

ONE

The First Science Texts

*The first written attempt to explain the physical world
in physical terms*

Life is short, and Art long, the crisis fleeting; experience
perilous, and decision difficult.
        —The *Aphorisms* of Hippocrates, ca. 420 BC

Hippocrates, the Greek doctor, lived in a world of solids and gods.

The solids surrounded him. The grey-green leaves of the olive trees, the earth beneath his feet, the brains and bladders of his patients, even the wine that he drank (in moderation); all of these were absolute, uncompounded, simple. How they came to be in their present forms, how those forms might change in the future—these questions occupied Greek scholars for long hours. But what *composed* them, what intricacies might lie beneath their surfaces and *explain* them; asking this was like interrogating a rock.

Twenty-three centuries later, Albert Einstein and the physicist Leopold Infeld jointly offered an analogy for the Greek plight. The ancient investigator of the natural world was like

> a man trying to understand the mechanism of a closed watch. He sees the face and the moving hands, even hears its ticking, but he has no way of opening the case. If he is ingenious he may form some picture of a mechanism which could be responsible for all the things he observes, but he . . . will never be able to compare his picture with the real mechanism and he cannot even imagine the possibility or the meaning of such a comparison.[1]

Instead of mechanisms, the Greeks had gods.

The gods lived among the solids of the natural world, wandered through the olive groves, resided in their sanctuaries and shrines. They were always watching, judging, and warning men. "The gods . . . notice all my doings," explains a character in Xenophon's *Symposium*, "and because they know how every one of these things will turn out, they give me signs, sending as messengers sayings and dreams and omens about what I ought to do." The divine suffused and guided the natural order. "All things are full of gods," the mathematician Thales remarked, 150 years before Hippocrates: all things and all places.[2]

The Greeks studied, and philosophized about, both the presence of the gods and the properties of solid nature. They were curious, not blindly accepting. But their world was not divided into the theological and the material, as ours is. The divine and the natural mingled freely.

In this they were like their contemporaries. The Egyptians, who had honed astronomical observations to an exactness, had already constructed a calendar that accounted for the flooding of the Nile. They could predict when the star Sirius would begin to appear in the predawn sky just before the sun ("heliacal rising"), and they knew that Sirius's rising meant the inundation was on its way. Yet the certainty of their calculations didn't destroy their conviction that the Nile rose at Osiris's pleasure.[3]

East of Athens, Persian astronomers were tracking lunar and solar eclipses, hard on the trail of a new discovery: the saros cycle, a period of 6,585.32 days during which a regular pattern of eclipses plays itself out and then begins again. Their equations made it possible to forecast the next lunar eclipse with mathematical precision, which meant that the temple priests had enough time to prepare rituals against the evil forces that a lunar eclipse might release. (According to Persian documents from about 550 BC, precautions involved beating a copper kettledrum at the city gates and yelling, "Eclipse!")[4]

For the Greeks, too, *supernatural* and *natural* existed in the same space. In fact, it was the god-believing mathematician Thales who came up with what may be the first scientific theory: Despite its solid appearance, the entire universe is made of water. His writings on the subject are long lost, but Aristotle preserved their argument in the *Metaphysics*, 300 years later.

> Thales . . . stated [the foundational principle of the universe] to be water. (This is why he declared that the earth rests on water.) Perhaps he got this idea from seeing that the nourishment of all things is moist . . . and also because the seeds of all things have a moist nature; and water is the principle of the nature of most things.[5]

Water (as it turned out) was the wrong explanation. But Thales's theorizing is the earliest known attempt to peer inside the universal watchcase and see what else, independent of divine power, might be causing it to tick.

Thales's attempt to discover an underlying truth about the universe without reference to the gods ("Thales's Leap," biologist Lewis Wolpert calls it) was probably not the first Greek theory of its kind, but it is the first preserved by name. Thales's actual works have disappeared, though. Thales's Leap may be the first known scientific theory, but the Hippocratic *Corpus*—a collection of some sixty medical texts that explain disease without blaming or invoking the gods—is the first surviving book of science.

The entire *Corpus* was once attributed to the shadowy Hippocrates himself, a fifth-century doctor who grew up on the tiny Greek island of Kos, just off the coast of Asia Minor. Plato tells us that Hippocrates taught aspiring doctors for a fee; the *Corpus*, now generally accepted as a collective project of his students and followers, preserves his lessons.[6]

Many of Hippocrates's contemporaries were priest-physicians, devotees of Aesculapius (son of Apollo, god of healing). To be cured by Aesculapius, a patient would travel to one of the temples of the god and spend the night in the *abaton*, the temple's sacred dormitory, surrounded by the free-slithering snakes that represented the god's presence. Sometime during the night, healing would take place. The serpents would lick the patient's wounds and mend them, or the god would send a dream explaining how the illness should be treated. Or perhaps Aesculapius himself would appear to carry out the cure. "Gorgias of Heraclea had been wounded with an arrow in one of his lungs," writes the Greek chronicler Pausanius:

> Within eighteen months the wound generated so much pus that sixty-seven cups were filled with it. He slept in the dormitory, and in a dream it seemed to him that the god removed the barb of the arrow with his lung. In the morning he went forth whole, with the barb of the arrow in his hands.[7]

Hippocrates didn't necessarily disbelieve in Aesculapius's existence, but he was skeptical about the god's role in illness. Instead, he looked to the visible world, the ordered cosmos, for explanations. Diseases were not caused by angry deities, and they did not need to be cured by a benevolent one. Even epilepsy, long held to be a sacred condition inflicted by demons or divine possession, was "no more divine nor more sacred than other diseases, but has a natural cause." The only reason to chalk up illness to the will of a god is ignorance: "This notion of its divinity," Hippocrates says tartly, "is kept up by men's inability to comprehend it."[8]

Hippocrates blamed stomach upsets, fevers, epilepsy, plagues, and illnesses of all kinds on imbalance—too much or too little of one of the four fluids, or "humors," that course through the human body. When these four fluids (bile, black bile, phlegm, and blood) exist in their proper proportions, the body is healthy. But any number of natural factors might throw them out of whack. For Hippocrates, the chief causes of unbalanced humors were winds (hot winds, for example, caused the body to produce far too much phlegm) and water (drinking stagnant water could lead to an overabundance of black bile). The recommended treatment: restore the body's balance. Purges and bleeds were prescribed to get rid of excess humors. Herbs (rue, mustard, fennel, stinging nettle) helped to draw out some humors and renew others. Sick men and women were often sent to different climates, away from the winds and waters that were deranging their natural harmonies.[9]

Like Thales's theories, Hippocrates's explanations were wrong.

But half the time, entirely by accident, his methods worked. Avoiding marshy, stagnant water *did* improve health. Shifting from a crowded, epidemic-afflicted city to a breezy coastal town *could* bring recovery from an illness. Light nutritious meals *were* helpful to feverish patients—much more helpful than a long onerous journey to the nearest *abaton* and an uncomfortable night spent with snakes.

While the temples of Aesculapius didn't immediately go out of business, Hippocratic methods slowly gained traction—so much traction that in the eighteenth century, physicians were still purging, bleeding, and sending their patients to the seaside. The Hippocratic worldview even lingers today. I know perfectly well that a cold is a viral infection, but I still find myself yelling, "Don't go outside without a coat or you'll catch cold!" as my sons bound into a windy winter morning clad only in T-shirts and shorts.

Thus the Hippocratic *Corpus* stands not only as the first surviving scientific writing, but the first recorded triumph of natural methods over the unearthly.

To read relevant excerpts from The Corpus, *visit http://susanwisebauer .com/story-of-science.*

HIPPOCRATES
(ca. 460–370 BC)
On Airs, Waters, and Places

The nineteenth-century Francis Adams translation, one of the first done for English-speaking lay readers, is still readable and is available both in print and as an e-book. The Adams translation includes *On Airs, Waters, and Places*, along with the *Aphorisms*, *The Oath of Hippocrates*, and several other works, collected together as *The Corpus*. Editions include

The Corpus, Kessinger Legacy Reprint (paperback, 2004, ISBN 978-1419107290).
The Corpus, Library of Alexandria (e-book).
The Corpus, with foreword by Conrad Fischer, Kaplan Classics of Medicine (e-book and paperback).

The Adams translation of *On Airs, Waters, and Places* alone is available online in multiple places.

A more modern translation is included in the Penguin Classics paperback *Hippocratic Writings*, translated by John Chadwick and W. N. Mann, with an introduction by G. E. R. Lloyd (paperback, 1983, ISBN 978-0140444513). The sentence structure is slightly easier to follow, but the two translations are very similar.

Beyond Man

The first big-picture accounts of the universe

Everything consists of the atoms . . . and there is nothing else.
                                    —Plutarch, on Democritus

All men, Socrates, who have any degree of right feeling do
this at the beginning of every enterprise great or small—
they always call upon the gods.
                                    —Plato, *Timaeus*, ca. 360 BC

Hippocrates and his followers were, without doubt, doing science.
They looked to natural factors to explain the natural world, and their
calling required more study than piety, more knowledge than faith.

But they still had no capacity to peer inside the mechanism.

The four humors had never been seen. The effects of waters and
winds upon them had never been tested or proved. The human
body still presented an impermeable surface to its examiners; the
Greek physicians could describe its anatomy reasonably well, but
the processes that went on within it were sealed away from their
view. So instead of looking more closely,* Hippocrates and his followers tried to *reason* their way in. Their favorite analytical tool was

---

*Dissection was not practiced widely, most likely because of the ancient Greek
belief that proper burial was the door into a satisfactory afterlife. James Longrigg
provides a useful overview of the Greek attitudes toward dead bodies in *Greek
Rational Medicine: Philosophy and Medicine from Alcmaeon to the Alexandrians* (Routledge, 1993), 184ff.

the analogy: The eye is like a lantern, so it must contain fire. The internal organs are like copper vessels that hold humors; therefore they must be connected by a system of tubing that allows humors to shift from one organ to another.[1]

This method of doing science had almost nothing to do with observation. In the absence of scientific tools, without a shared scientific vocabulary, lacking the most basic agreement over the foundational principles of the universe, the Greeks created elaborate hypothetical structures and then fitted symptoms carefully into them. Upset stomach? The tubing between your internal organs must be blocked, creating a buildup of humors. Treatment: flush the tubing with a purge, just as a plumber would flush a blocked pipe.

The Hippocratic doctors, characterized by the professional myopia that still afflicts many physicians, did not look very far beyond the body's immediate surroundings. But in the centuries after Hippocrates, Greek thinkers began to speak and write in similar ways (naturalistic ways, driven by analogies) about *phusis*: the ordered universe, the whole natural realm.

Phusis, often translated simply as "nature," encompassed much more than the natural world as we would now think of it. The ordered universe consisted of the earth and men. To study *phusis* was to study both politics and plants, the soul and the stars. The Greeks inhabited a fluid and unbounded intellectual landscape; speculations about the composition of the sea and sky mingled seamlessly with political philosophy.[2]

Lacking any tradition of close observation, in the absence of any scientific tools that would allow them to pick apart a phenomenon into its component parts, the Greek thinkers instead attempted to explain *phusis* in its entirety, from origin to its present form. They were armchair time travelers, creating elaborate story lines for the universe. The monists believed that it all began with a single underlying element, one sort of *stuff*, containing within itself the principle of its own change: "This, they say, is the element and this the principle of things," Aristotle wrote much later of the monists, "some entity . . . from which all other things come to be." Thales, one of the earliest monists, had proposed water; the sixth-century philosophers Anaximenes and Heraclitus suggested air and fire, respectively; Anaximander, around 575 BC, proposed something called "the indefinite," a thing that has itself no characteristics but contains opposite qualities that separate, producing change. The pluralists, on the other hand, were in favor of multiple underlying elements; Empedocles, around 460 BC, suggested four—earth, air, fire, and water—an arrangement widely adopted by other thinkers.[3]

And then there were the atomists, most notably the shadowy Leucippus and his much-better-known pupil Democritus, both of them teaching and writing in the last quarter of the fifth century. "Leucippus . . . posited limitless and eternally moving elements, the atoms," the philosopher Simplicius explains. Democritus expanded on his master's theory; these atoms "are so small that they escape our senses. . . . From them, as from elements . . . the visible and perceptible masses" are formed.[4]

As it turned out, this was more or less true.

The atomists are often celebrated as eerily farsighted and discerning. Actually, they were no more gifted than the monists or pluralists; it just so happened that, like Hippocrates, they accidentally hit on some elements of the truth. "These early atomists may seem wonderfully precocious," remarks physicist (and Nobel laureate) Steven Weinberg, "but it does not seem to me very important that the [monists] were 'wrong' and that the atomic theory of Democritus and Leucippus was in some sense 'right.' . . . How far do we progress toward understanding why nature is the way it is if Thales or Democritus tells us that a stone is made of water or atoms, when we still do not know how to calculate its density or hardness or electrical conductivity?"[5]

In other words, the watchcase was still firmly closed; these early science writers were theorizing with no way to check their results. And we cannot even read their actual words, because their texts, like Thales's writings, have all been lost. Their speculations are preserved only in the summaries and studies written by others, long after the fact. Sextus Empiricus, who gives a detailed summary of Democritus's teachings in his work *Against the Mathematicians*, lived six hundred years later; Simplicius, one of the few to quote Leucippus directly, was born a millennium after his subject.

But together, the monists, pluralists, and atomists* took a critical step forward. All of them affirmed the same principle: *Phusis*, like human illness, could be explained in purely material terms. If the atomists stand a little to the fore, it is only because Democritus was even more insistent than his colleagues that the universe consisted of nothing but atoms and what he called "the empty"—the place in which atoms rushed about, collided and intertwined by chance, and separated by coincidence. There were gods in Democritus's world, but they too were made up of atoms; they too were subject to the laws of nature; they created nothing, and they too would eventually be dissolved. There was no plan. There was no design. There were simply atoms, moving randomly in the empty.

Democritus cast the gods out of beginnings. His explanations for the existence of the universe were all materialistic; and like materialistic explanations ever since, they inspired vigorous opposition.[6]

The most vigorous, and the most influential, came two generations after Democritus, from the long-lived Athenian philosopher Plato. Without the gods, objected Plato, ethics were doomed. Without godly origin, the state would disintegrate. Without supernatural creation, human morality would vanish. Therefore, *phusis* might be comprehended by the senses, but its beginnings must be explained with reference to the divine.[7]

So in the *Timaeus*, written late in his life, Plato offered his own sketch of the universe and how it works—the first self-consciously big-picture neoscientific treatise, the first known attempt to offer a theory of everything. It is a hybrid work, beginning with the origins of the universe at the hands of a divine creator—a divine force, an unknown Craftsman—and then moving from origins to an explanation of the universe's present function that has *no* reference to the divine. Plato lived in a world where it was no longer possible to ascribe the rising of rivers and the motions of

---

*They are often jointly known as the "pre-Socratic philosophers," a misleading term because it covers both the philosophers who came before Socrates (ca. 469–399 BC) and those who came after him but who disagreed with the Platonic point of view.

the moon to the will of gods. Yet he could not imagine a universe that had always been, or a beginning that was not sparked by the divine.

Looking around at the ordered world, Plato sees design and beauty; this, he reasons, must come from a mind, from an intelligence. So he begins with an account of a Demiurge (a being who fashions what we see from materials that already exist) and the shaping of a good, spherical universe out of disorderly, irregular matter. This good universe exists first, uncorrupted and perfect, in the mind of the Demiurge; as it comes into physical being, it slips very slightly away from this Ideal, taking shape as a visible and inferior Copy, a physical shadow of the original Reality.

This physical universe consists of four elements—earth and fire the most elemental, with air and water as a bond between them. Human beings, living in the corporeal universe and in some ways mirroring its structure (as water circulates through the earth, so blood circulates through the body), can understand it through the senses: touch, smell, sight, hearing.

And in this perception, the Craftsman has no part.[8]

This is, of course, a simplification of the *Timaeus*. "Of all the writings of Plato, the Timaeus is the most obscure and repulsive to the modern reader," begins the introduction to Benjamin Jowett's classic translation of Plato's *Dialogues*, and he does not overstate the dialogue's complexity. Striving to describe physical phenomena in a language that had been devoted to poetry and philosophy, Plato writes so obliquely that it is often a struggle to figure out the topic of a given sentence, let alone its conclusion. And interwoven with his descriptions of how we sense and interpret are long discourses on philosophical distinctions (for example, between *what always is and never becomes* and *what becomes and never is*), a bizarre explanation of why the universe doesn't have feet but we do, and the story of the lost civilization of Atlantis.

Yet the *Timaeus* affected the practice of Western science for the next two thousand years. Plato divided origins from observation, creation from the explanations of everyday phenomena. He acknowledged the importance of the senses in the study of the world around us—and, like Hippocrates, opened up an ever-widening space in which science could be practiced without appeal to the supernatural.

This was a great gift to the newborn field of science. But Plato's bequest carried with it a fatal infection. Yes, man can understand the physical world through the senses. But since physical reality is a shadow of the ideal, physics is always subordinate to metaphysics. Philosophy struggles to understand the Ideal from which the world has declined. Science merely uses observation to understand the declined shadow itself. And so science can never lead to truth; it must always sit at philosophy's feet, willing to receive correction.

To read relevant excerpts from the Timaeus, *visit http://susanwisebauer .com/story-of-science.*

PLATO
Timaeus
(ca. 360 BC)

Benjamin Jowett's nineteenth-century translation is still widely reprinted and is clear and accessible to modern readers. Although it is not included in all editions of Plato's collected *Dialogues*, it can be found in

The Dialogues of Plato in Four Volumes, vol. 2, Charles Scribner's
    Sons (e-book, 1892).
*Dialogues of Plato: Translated into English with Analyses and Introduc-
    tion*, Cambridge University Press (paperback, 2010, ISBN 978-
    1108012102).

A modern translation can be found in

Peter Kalkavage, trans., *Plato's Timaeus*, Focus Publishing (paper-
    back, 2001, ISBN 978-1585100071).

Unlike the Greek of Hippocrates, which deals with things (water, phlegm, purges, bellies), Plato's Greek consists largely of abstract terms that cannot easily be translated by single words. He deals with matters of being and existence, not diagnosis and pre-scription; his vocabulary is obscure and/or archaic; and on top of that, he is fond of wordplay, sly linguistic jokes, and puns. As a result, translations differ markedly. Here is Jowett's rendering of a passage from the first part of the *Timaeus*:

> Why did the Creator make the world? He was good, and desired that all things should be like himself. Wherefore he set in order the visible world, which he found in disorder.[9]

And here is Peter Kalkavage's translation of the same passage:

> Now let us say through what cause the constructor constructed becoming and this all. Good was he, and in one who is good there never arises about anything whatsoever any grudge, and so, being free of this, he willed that all things should come to resemble himself as much as possible. That this above all is the lordliest principle of becoming and cosmos one must receive, and correctly so, from prudent men. For since he wanted all things to be good and, to the best of his power, nothing to be shoddy, the god thus took over all that was visible, and, since it did not keep its peace but moved unmusically and without order, he brought it into order from disorder, since he regarded the former to be in all ways better than the latter.[10]

Kalkavage's translation is carefully literal, unpacking each Greek construction and largely refraining from interpretation; Jowett's combines translation with an explanation of Plato's meaning, and so is easier for the nonspecialist to grasp.

If you're anxious to understand the full dimensions of the philo-sophical problems, choose Kalkavage. If you simply want to get a general sense of Platonic idealism as it affected the practice of sci-ence over the next thousand years, stick with Jowett.

THREE

Change

The first theory of evolution

Everything which is altered is altered by things which are perceptible to the senses.

—Aristotle, *Physics*, Book VII, ca. 330 BC

Nature passes so gradually from inanimate to animate things that the boundary between them is indistinct.

—Aristotle, *History of Animals*, Book VIII

Like his teacher Plato, Aristotle saw beauty and order around him. But while Plato saw this beauty as proof of a Craftsman at the beginning, Aristotle saw it as a signpost pointing toward fulfillment at the end.

This altered everything—most especially, change.

For Plato, change does not mean progress. Only decay. His natural world is an inferior copy of the Craftsman's original concept, which means that it is an inherently flawed work of art—like a perfectly written play that inevitably accumulates myriad minor defects as soon as it is staged by real actors, in real costumes, wandering through real scenery. The physical world is always *less* than it was meant to be, and any change inevitably pulls it further and further away from the ideal.

But Aristotle, watching a sprout grow into a tree, a cub into a lion, an infant into a man, saw something else.

First, he wanted an explanation of the process: How do these changes *happen*? In what stages does one entity, one *being*, assume more than one form? What impels the change, and what determines its ending point?

Then, he wanted a reason. *Why* does a kitten become a cat, a seed a flower? What impels it to begin the long journey of transformation? Why is the state of kittenness, the existence of a seed in itself, not enough?

Today, when the cellular changes of growth are common knowledge, when every kindergarten class sprouts a bean on damp cotton, these questions seem superfluous. But part of the genius of Aristotle (wrongheaded though his conclusions often were) was to *ask* them. He did not assume that growth and change, as natural processes, were simply to be accepted. It was *because* they were natural processes that he questioned them; it was *because* they occurred as part of the natural cycle that he hoped to understand them.

This was science.

And so Aristotle's most seminal scientific work, the *Physics*, is all about change. Change in every sense: the natural change that happens when a young creature ages; the change that happens when an object moves from one place to another; whether the object is the same in the latter place as it was in the former; the explanation for why movement takes place at all.

The answer to this last is not, as it was for Plato, *decay*. It is, rather, that each object and being in the natural world *must* move from its present state into a future, more perfect one. Built into the very fabric of the seed, the kitten, the infant, is the potential for *change*: the "principle of motion," tracking steadily toward a glorious fulfillment. Aristotle's natural world, the *phusis* around him, is not a play that has declined from its ideal imagining. It is a documentary drama, moving toward a satisfying conclusion.

The results on the practice of science were dramatic.

For one thing, only Aristotle's point of view makes empirical inquiry—the observation and understanding of the physical world—a true path to real knowledge, *valuable* knowledge. Platonic thought, always casting the physical world as inferior to the Ideal, inevitably devalued scientific study, making it a lower-order, second-rate enterprise. (George Sarton, the chemist-historian who founded the discipline of the history of science, went so far as to call the influence of the *Timaeus* on later scientific thinking "enormous and essentially evil.") But in Aristotle's philosophy, science leads to truth.[1]

Furthermore, only Aristotle's thought makes evolution possible.

In Plato's world, change is corruption, and movement is always *away* from the ideal, toward a less effective, less developed state. But in Aristotelian science, nature is developing inexorably toward a more fully realized end.

This is not exactly what we mean by evolution today. Biological evolution has no predetermined goal, no overall design. Aristotle's science is *teleological*, firmly convinced that nature is moving purposefully, in the direction of perfection.

Teleology can be an expression of faith in a Creator, a God who has set the world on a certain course that cannot be avoided. But Aristotle did not see nature as the creation of a divine Designer. He did not break from his master so far as to do away with a First Cause; at the beginning of all movement, Aristotle surmised, must be the original Force, that which cannot be worked on by any other cause, the Unmoved Mover. But this Unmoved Mover does not shape nature from the outside. It is no potter, and the world is not clay. A sprout becomes a tree because its *treeness* is already inherent inside it. For Aristotle, teleology is not an external guiding force, but an internal potentiality. The key to nature's final form is already *inside* it.

The philosophical basis for what Darwin would later articulate had been laid.

Aristotle's vision of a world where movement is always forward, always *purposeful* (never purely random, as Democritus and the atomists had proposed) served as the framework for what he called the *scala naturae*, the Scale of Nature: a graded, continuous ranking of natural organisms from the simplest to the most complex. Plants were at the bottom, human beings nearer the top. In the hands of medieval philosophers, Aristotle's Scale of Nature would become the Great Chain of Being, a connected ranking of all natural elements and beings. Rock sat on the bottom rung, God at the top.[2]

To locate living organisms on the Scale of Nature, the natural scientist had to understand them. Aristotle's other great scientific works, his natural-history compendiums (*History of Animals, Generation of Animals, Parts of Animals*), describe, organize, and classify living things, discovering within unwieldy nature itself the principles of purposeful change laid out, in the abstract, in the *Physics*.

This task was complicated by the total lack of Greek terms for such an enterprise. Doing science before the language of science had been created, Aristotle had to make up his own vocabulary, his own terms and titles and divisions, as he went along. In doing so, he invented taxonomy: the science of grouping living things together by their shared characteristics. His most essential division—between bloody and bloodless animals, the red-blooded and the nonsanguinous—still exists today, restated as the separation between vertebrates and invertebrates.[3]

He also lent to later science plenty of whopping errors: spontaneous generation, a spherical universe made up of rotating crystalline shells, and the inheritance of acquired traits (even wounds and blows). But even wrong—sometimes perversely wrong—his theories were always based on his own observations of natural change. He had rescued change from Plato's dust heap and elevated it into the central principle of nature: an engine that would drive scientific study inexorably forward.

To read relevant excerpts from the History of Animals *and the* Physics, *visit http://susanwisebauer.com/story-of-science.*

ARISTOTLE
(384–322 BC)
History of Animals

Richard Cresswell's still-readable nineteenth-century translation is available as a free e-book.

Johann Gottlob Schneider, ed., *Aristotle's History of Animals in Ten Books*, trans. Richard Cresswell, Henry G. Bohn (e-book, 1862).

The best print version is the much more expensive Loeb Classical Library version in three volumes.

Aristotle and A. L. Peck, *Aristotle: History of Animals, Books I–III* (Loeb Classical Library no. 437), Harvard University Press (hardcover, 1965, ISBN 978-0674994812).

Aristotle and A. L. Peck, *Aristotle: History of Animals, Books IV–VI* (Loeb Classical Library no. 438), Harvard University Press (hardcover, 1970, ISBN 978-0674994829).

Aristotle, Allan Gotthelf, and D. M. Balme, *Aristotle: History of Animals, Books VII–X* (Loeb Classical Library no. 439), Harvard University Press (hardcover, 1991, ISBN 978-0674994836).

ARISTOTLE
(384–322 BC)
Physics

The R. P. Hardie and R. K. Gaye translation, done as part of a forty-year effort to translate Aristotle into a standard English version (the "Oxford Translation"), is available as a free e-book or as a paperback reprint.

Aristotle, *Physics*, trans. R. P. Hardie and R. K. Gaye, Clarendon Press (1930, e-book at "The Internet Classics Archive" online, paperback reprint by Digireads, ISBN 978-1420927467).

A more recent and very readable translation by Robin Waterfield is worth investigating if you intend to read all eight books of the *Physics*.

Aristotle, *Physics*, trans. Robin Waterfield, Oxford World's Classics, Oxford University Press (paperback, 1999, ISBN 978-0192835864).

FOUR

Grains of Sand

The first use of mathematics to measure the universe

These things will appear incredible to the great majority of
people who have not studied mathematics, but . . . to those
who are conversant therewith . . . the proof will carry
conviction.
    —Archimedes, "The Sand-Reckoner," ca. 250 BC

Up until this point, the brand-new field of science had made almost
no use of mathematics. Greek mathematics had been following its
own separate, somewhat winding road; and there had, as yet, been
no major crossroads with the study of nature.

The god-believing mathematician Thales was credited with first
coming up with the formulation of abstract, universal mathematical laws. The Greeks were not the only ancient people who knew
their geometry—mathematicians in the Indus valley were already
there—but before Thales, we have no record of any thinker going
beyond specific geometric observations ("A circle is bisected by
any of its diameters") to proofs showing that those observations are
always true, for all circles, everywhere in the universe.[1]

Since Thales, geometry—the study of angles and lengths, the
areas they create and the patterns they follow—had developed
into the root and trunk of Greek mathematics. Arithmetic (the
branch of mathematics dealing with numerals) was derived from it.
Numerals were, first and foremost, ways of measuring geometric

properties such as area and length. And those measurements were usually expressed not simply as numbers, but as ratios.

In other words, asked the measurements of this rectangle:

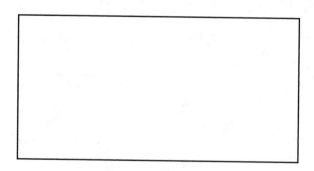

we would most naturally label the long sides as 3 inches in length, the short sides as 1½ inches each. But the Greek mathematician would express this measurement as the relationship between the two sides:

$$2{:}1$$

because the relationship of the sides to each other is the same as the relationship of the numeral 2 to the numeral 1.

The Greeks could use ratios to add, multiply, and perform all of the other operations you learned to carry out in arithmetic class. It is for this reason that mathematics still speaks of *rational numbers*: any number that expresses the ratio between two integers is rational.*

After Thales and before Plato, the most active mathematical work was done by the Pythagoreans: followers of Pythagoras, a Greek mystic who lived sometime in the sixth century BC, and about whom practically nothing is known. What details of his life survive come entirely through his later disciples, such as Iamblichus, who lived some eight hundred years after the master and took as his life's work a ten-volume encyclopedia of Pythagoras's

---

* For example, the fraction ⁴/₉, because it is the ratio of 4 to 9 (or, 4 divided by 9); or 71, because it is the ratio of 71 to 1 (or, 71 divided by 1); or –11, because it is the ratio of –11 to 1 (–11 divided by 1). Another way to look at it: a rational number can always be put into the form a/b, as long as a and b are both integers and b does not equal zero.

teachings. Pythagoras, says Iamblichus, was descended on both his mother's and father's sides of the family from Zeus, and it was rumored that the boy himself was the son of Apollo, who visited his mother while his father was away. (This, Iamblichus admits, is "by no means" certain, but "no one will deny that the soul of Pythagoras was sent to mankind from Apollo's domain.")[2]

Pythagoras was venerated as a divine mouthpiece, and his mathematics was primarily *not* a tool to understand the natural world. It was a method of understanding truth itself. Mathematics, Pythagoras taught, was the *only* path to knowledge: without numbers, nothing could be truly apprehended. Numerals had oracular power—particularly 1, 2, 3, and 4, which could be joined to create all existing dimensions. The sum of these four numbers, 10, was a holy number, the *tetractys*.[3]

Pythagoreans were vegetarians and teetotalers. They believed in the transmigration of souls, practiced shrouded black rites, and taught that the intervals of musical notes revealed deep truths about the universe (a theory that developed, long after, into the medieval Harmony of the Spheres). But interwoven with the cabalism was some authentically meticulous mathematics. The Pythagorean theorem, the first geometric idea encountered by most seventh-graders, had long been known to ancient mathematicians (the Egyptians certainly understood it), but the Pythagoreans first phrased it as a universal law, a truth that applied to *all* right-angled triangles everywhere.[4]

This theorem seems to have led the Pythagoreans to realize, apparently for the first time in recorded history, that there were such things as irrational numbers.

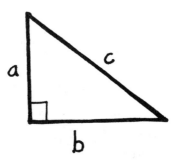

4.1 THE PYTHAGOREAN THEOREM: $a^2 + b^2 = c^2$

Later writers chalked the discovery up to the Pythagorean philosopher Hippasus, who lived just before 400 BC. Hippasus, working on triangles, realized that no single unit could be used to measure both *c* and *a*. The two lines are "incommensurable," having no common measure. In other words, there is no way to express the relationship between them using the numerals that the Pythagoreans had available to them; it is impossible, for example, to say that *a* is ⅓ or ¼ of *c*. The discovery of incommensurables in geometry led shortly after to their arithmetical parallel, *irrational numbers*—numbers that cannot be expressed in terms of the ratio of one whole number to another.

Apparently this was a staggering blow to Pythagorean mysticism, since the entire system was built around the conviction that all natural relationships could be expressed using ratios. Hippasus's discovery was so disturbing that it brought down the ire of the universe on his unfortunate head: "It is well known," remarks a later commentary, "that the man who first made public the theory of irrationals perished in a shipwreck in order that the inexpressible and unimaginable should ever remain veiled."[5]

A mathematical tradition that shies so strongly away from its own implications in the real world is a mathematical tradition that probably won't be useful in science; and for the first centuries of its existence, Pythagorean mathematics remained both shielded from the uninitiated and separate from the slowly developing new field of science. It was a religion, not a science, designed almost entirely for the contemplation of the divine, rather than the study of the earth.

•

Neither the Aristotelian nor the Platonic tradition made much use of mathematics in scientific writings, for completely different reasons.

Aristotle wanted to know what things were in themselves, not what their measurements were. Weight, height, circumference, diameter: all of these things change, and none of them, for Aristotle, point to what a natural object *is*. Mathematics gave him no insight into the *plantness* of a plant, the waterness of water. His natural classifications, like his physics, made use of qualities, not quantities; the presence of blood rather than the size of the heart, the habits of the burrower rather than the measurements of the burrow. Mathematics, for Aristotle, was irrelevant.

For Plato, mathematics was essential—as long as it wasn't corrupted by contact with the physical world.

He explains this most clearly in Book VII of the *Republic*, which famously contains the "Allegory of the Cave." There is a difference, Plato insists, between the uncorrupted and perfect universe, the Ideal as it exists in the mind of the Demiurge, and the visible and inferior Copy in which we live. The unphilosophical man sees only the Copy. He is like a prisoner chained in a cave, able to see only the shadows of the reality outside reflected on the cave wall, while the reality itself remains blocked from his eyes. Should the prisoner be suddenly unchained and brought into the full sunlight of the world outside the cave, the glare would be so painful and dazzling that he would be unable to see clearly. He would willingly return to captivity in the cave, preferring to gaze at the shadows rather than reality.

So the unphilosophical man must be brought slowly out of the "cave" into which he was born, into the full sunlight of philosophical knowledge. He must be carefully taught to *understand* the Ideal, so that he will no longer be satisfied with the Copy. Arithmetic and geometry are tools of the transformation. Arithmetic, because it begins by distinguishing between the concepts of *singular* and *many*, unity and plurality, the one and the infinite, has "a power of drawing and converting the mind to the contemplation of true being." Geometry, likewise, leads the soul toward truth, and creates the "spirit of philosophy" where none had before existed.[6]

But both arithmetic and geometry have this power only when they deal with abstractions. The study of arithmetic, Plato cautions, must be "pursued in the spirit of a philosopher, and not of a shopkeeper . . . arithmetic has a very great and elevating effect, compelling the soul to reason about abstract numbers, and rebelling against the introduction of visible or tangible objects into the argument." Geometry must be practiced, not as most geometricians do ("They have in view practice only, and are always speaking, in a narrow and ridiculous manner, of squaring and extending and applying and the like—they confuse the necessities of geometry with those of daily life"), but as a method of understanding Ideal forms and shapes. "The knowledge at which geometry aims," he explains, "is knowledge of the eternal, and not of aught perishing and transient."

In fact, because Plato's philosophy dictated that reason, untainted by the input of the senses, can yield truth, Plato was suspicious of any arithmetical conclusions that were related to observation. This made him wary, for example, of astronomy. Astronomers observed past movements of heavenly bodies, analyzed them, and used mathematics to calculate their future positions. But such calculation incorporated observation—the senses—into the practice of mathematics, thus bringing it from the realm of the ideal into the realm of the shadow. So, while Plato acknowledged the value of astronomical calculations, he also warned astronomers not to assume that their theories actually *described* the universe. The astronomer's conclusions might be *likely*, but in no way could they be characterized as *true*.[7]

This dismissal of the real world still echoes down to us in the language used by math departments; applied mathematics is no longer scorned as "shopkeeping," but theoreticians still claim the snobby title of "pure mathematics" for their own discipline.[8]

·

Despite Plato's scorn, mathematicians were, increasingly, trying to find the intersection between maths and natural science.

Indeed, it seems likely that Plato's snarky remark about "shopkeeper" mathematics was directed at one of his contemporaries, a Pythagorean mathematician named Archytas who dared to depart from the mysticism of his colleagues in order to apply mathematics to real problems in the real world. The third-century biographer Diogenes Laertius calls Archytas "the first who methodically applied the principles of mathematics to mechanics." Tradition held that Archytas had invented a wood dove that actually flew. And in his *Politics*, Aristotle remarks offhand that since young children are unable to be quiet, they should be given a toy rattle designed by

Archytas, to occupy them and "stop them from breaking things in the house." But only fragments of Archytas's own writings remain, and the scraps shed no light on his scientific pursuits.[9]

Not until the middle of the third century did a piece of scientific writing make use of mathematics for its investigation—and survive. The text is called "The Sand-Reckoner," and it was written by Archimedes, a native of the Sicilian city of Syracuse.[10]

Archimedes had an advantage that Archytas lacked: a handbook, written by Euclid (ca. 325–265 BC), assembling the geometric knowledge that had been floating around Pythagorean circles for centuries into thirteen nonmystical, entirely unreligious books. The *Elements* begins with a list of definitions ("A point is that which has no part. . . . An obtuse angle is an angle greater than a right angle") and continues with what Euclid calls "postulates" and "common notions"—both being statements so self-evident that they do not need to be proven. Postulates have to do with geometry in particular ("All right angles are equal to one another"), and common notions apply to both geometry and other fields of study ("The whole is greater than the part").

But the meat of Euclid's book is in his geometric proofs—problems and solutions showing that the rules of geometry work in all places, for all times, for everyone. There is nothing less mystical and obscure than Euclid's proofs. Given his starting assumptions, and perhaps a ruler and compass, everyone (not just the initiates) could comprehend his system.

The fifth-century Greek philosopher Proclus, who wrote an extensive commentary on Euclid's *Elements* some seven hundred years later, relays a well-known story about Euclid and Ptolemy I, the king of Egypt. Ptolemy, an ex-general of Alexander the Great, was no scholar. But he knew the value of learning and tackled the *Elements*, only to find it a little too meaty for his taste. So he asked Euclid for a simpler way to understand its principles. "Sir," Euclid responded, "there is no royal road to geometry."

The story may be apocryphal, but it reveals the new truth of geometry. There were no shortcuts, no divine revelations or ritual sacrifices required, and also no privilege. The *Elements* rescued geometry from the Pythagoreans and gave it to the world.[11]

Almost immediately afterward, Archimedes connected this new tool to the contemplation of the universe.

Archimedes is best remembered for a discovery that might not have happened. According to the Roman biographer Vitruvius, two hundred years later, Archimedes had been tasked by his king with the job of figuring out whether a dishonest goldsmith had stolen some of the gold intended to go into the royal crown and replaced it with cheaper silver. The crown *looked* gold, and it was the correct weight. But was it pure?

"While the case was still on his mind, he happened to go to the bath," writes Vitruvius, "and on getting into a tub observed that the more his body sank into it, the more water ran out over the tub. As this pointed out the way to explain the case in question, he jumped out of the tub and rushed home naked, crying with a loud voice that he had found what he was seeking; for as he ran he shouted repeatedly in Greek, 'Eureka, eureka' [I have found (it)]." He had just realized that, since silver is lighter than gold, a crown of pure gold would have to be slightly smaller than a crown of equal weight made up of a gold and silver combination. So, if the adulterated (and thus larger) crown were submerged into a jar of liquid, it would displace *more* water than the pure, microscopically *smaller* crown. Calculating how much water the unadulterated crown should have displaced, and comparing it with the actual displacement, Archimedes was able to measure the extent of the silver substitution, and *manifestum furtum redemptoris*: the stolen property was revealed.[12]

Vitruvius is notoriously unreliable, and more than one experimenter has pointed out that it would be almost impossible to measure the tiny difference in water displacement with enough accuracy to predict the exact makeup of the crown. But there is no question that Archimedes understood the science behind the story. In his own *On Floating Bodies*, he describes the Principle of Buoyancy ("Archimedes' Principle"), which is simply this: "A body partly or completely immersed in a fluid is lifted up by a force equal to the weight of the displaced fluid."[13]

Archimedes wrote several essays expanding on Euclidean geometry, and he is generally credited with a whole series of inventions, including Archimedes' screw (a pump used to raise water from a lower to a higher level), the ship shaker (a mechanical claw that could lift an attacking ship out of the water), a planetarium, and various sorts of levers. The proof for most of these is weak or nonexistent; Archimedes may well have improved them, but his own writings don't show that he invented them.

What they *do* show is a mathematician who knew how to apply his knowledge to scientific questions. In his essay "The Sand-Reckoner," Archimedes finally brought geometry to bear on the study of the natural world.

The premise of "The Sand-Reckoner" is fairly straightforward: How many grains of sand would it take to fill the universe? Although this may seem like a mere thought experiment, remember that the Greeks were accustomed to measuring in terms of ratios. Archimedes's question was not merely "How big is the universe?" It was, instead, "Is it possible to measure the universe using the mathematical tools that we possess?" *Ratio* was the tool he had in mind; and his quest was to discover whether a meaningful relationship existed between two natural objects of vastly different sizes: a speck of sand and the whole of physical reality.

For the purposes of his essay, Archimedes decided to adopt a less-than-widely-accepted model of the universe—one with the sun at its center. It was much more popular, in ancient times, to see the universe as a relatively compact set of interlocking spheres centered around the earth. But Archimedes thought that this smallish universe wouldn't set him much of a challenge.

Instead, he chose to work out his calculations as they would apply to another model of the universe, one proposed by his contemporary Aristarchus. "You are aware," "The Sand-Reckoner" begins, "that 'universe' is the name given by most astronomers to the sphere whose centre is the centre of the earth. . . . But Aristarchus of Samos brought out a book consisting of [the] hypotheses . . . that the fixed stars and the sun remain unmoved, that the earth revolves about the sun in the circumference of a circle, the sun lying in the middle of the orbit."[14]

According to Aristarchus, the stars were the equivalent of 100 million earth-diameters away from the center of the universe—

much, *much* farther away than the stars of the geocentric model. And to make the universe even *larger*, Archimedes decreed that "earth-diameter" (an unknown quantity) should be understood as 1 million stadia across.

One million stadia is almost 100,000 miles, which is way too huge; the earth is really only about 7,900 miles across, depending on which bulges you measure. Even so, Archimedes's calculations of the size of the universe yielded a ridiculously small number. His universe turned out to be about 10 trillion miles across, which we now know to be less than 2 light-years; a single light-year is about 6 trillion miles, and the Milky Way galaxy alone is perhaps 120,000 light-years in diameter.[15]

But the actual dimensions were beside the point. What mattered was the larger question: "Is it possible to use mathematical language to describe a reality that is bigger than anything we have been able to measure in the past?"

And to this, Archimedes was able to answer a resounding yes.

Expressing the answer was slightly complicated, though. Greek numbers couldn't count that many grains of sand. The largest number in the Greek system was the *myriad*, or 10,000, written as (capital mu). The myriad could be combined with other numbers: so, for example, since ε (epsilon) represented 5, the compound symbol Mε stood for 5 × 10,000, or 50,000.

The greatest quantity that could be expressed using this system was the myriad myriad, or the myriad times itself: 100 million (100,000,000). Archimedes needed more numbers. So when he reached 100,000,000, he designated the *myriad myriad* as a single number, written βM (β, beta, represented 2). This meant he could now write multiples of 100,000,000; 500,000,000, for example, was βMε.[16]

This was not the most elegant number system in the world (just by way of illustration, 785,609,574,104 had to be written as βM,ζωνς, αMʹνξ, δρδ),* but it allowed Archimedes to figure that it would take 10^{51} grains of sand to fill up the heliocentric universe

---

*My thanks to Russell Cottrell's online script "The Greek Number Converter" for converting Arabic to Greek numbers.

of Aristarchus. For the first time, a scientist had forced mathematics to serve the purposes of science, rather than the other way around. Instead of shaping the universe to the perfect, abstract, Ideal of mathematical knowledge, Archimedes had molded the language of mathematics so that it fit the reality of the universe.

And he had conveyed another very clear message as well. For centuries, sand had represented the uncountable; *like sand* and *like the stars* meant, very simply, "That which is beyond our numbering." Choosing sand as his measure of the star-filled sky, Archimedes had made a new assertion: There is nothing in the universe that cannot be counted, and understood, by man.

To read relevant excerpts from "The Sand-Reckoner," visit http://susan wisebauer.com/story-of-science.

ARCHIMEDES
(287–212 BC)
"The Sand-Reckoner"

The classic nineteenth-century translation by Thomas Heath is available as a free e-book.

Archimedes, *The Works of Archimedes*, trans. T. L. Heath, Cambridge University Press (e-book, 1897).

A print version published by Dover includes not only "The Sand-Reckoner" but also Heath's introductory essay and eight other short works of Archimedes, including "On the Sphere and Cylinder" and "Measurement of a Circle."

Archimedes, *The Works of Archimedes on Mathematics*, trans. Thomas L. Heath, Dover Publications (paperback, 2013, ISBN 978-0486420844).

Heath's translations transform Archimedes's Greek numeral system into exponential numbers readable by English speakers. In places, he retains the Greek in brackets.

FIVE

The Void

*The first treatise on nature
to dispense entirely with the divine*

All fails, and in one single moment dies.
— Lucretius, *On the Nature of Things*, ca. 60 BC

While Archimedes calculated, while Aristotle investigated change, while Plato taught of ideal Forms, the atomists continued to insist that physical reality is made up of nothing more than indivisible particles, traveling at random through the infinite void.

Democritus, who had acknowledged the existence of the gods but taught that they, too, were made of nothing but atoms and "the empty," had died around 400 BC; a later chronicler records that he was 104 and simply decided to stop eating. His disciples lived on, passing his maxims from one generation to the next. The most successful and notorious of them was Epicurus, founder of a philosophical school that began to meet in the garden of his Athenian home around 307 BC. Epicurus developed atomism into an entire philosophical system—one that would finally lend itself to the cause of science two centuries after his death.[1]

Like Thales, Aristarchus of Samos, and the early atomists, Epicurus did not leave us his writings. Only fragments of his works survive, along with a letter to the Greek historian Herodotus that seems to have been written, fortuitously, as a précis of his teachings: a deeply secular philosophy that saw no pattern or design in the world.

Nothing comes into existence from what does not exist . . . there is no thing outside the whole, there is nothing that could enter the whole and produce a change.

All things that exist are either bodies or empty space. . . . We call this empty space "void." . . . Besides these two kinds, bodies and empty space, it is impossible to conceive of anything else.

The building blocks of bodily natures are the atoms or indivisibles.

The atoms are perpetually in motion throughout the vast ages.

There is no absolute beginning of such motions, since both the atoms and the empty space have existed forever.[2]

Like Democritus, Epicurus saw around him only an "infinite and mechanical universe of interacting particles." The physical objects around us, he explained, came into being not by divine intervention, but because atoms—spinning through the void—sometimes give an unpredictable hop, a random jump sideways, slam into each other, and join up to create new objects.[3]

Epicurus wasn't particularly interested in the science of these atoms. He cared nothing for knowledge as knowledge; he had no appreciation for the beauty of a scientific theory, no satisfaction in understanding the workings of the natural world. His preoccupation was ethical: How, living in such a world, should men *act*? Given that there is no divine plan, no afterlife, no immortal soul, how then should we live? How does man reach *ataraxia*, peace of mind, when adrift in an uncaring and impersonal universe, with no navigator but chance, with no guarantee of a safe farther shore? "Remember," he wrote himself to his disciple Pythocles, "that, like everything else, knowledge of celestial phenomena . . . has no other end in view than peace of mind and firm convictions."[4]

This peace of mind was not reached, as Epicurus's enemies would later insist, through thoughtless indulgence in sensual pleasures. Rather, Epicurus struggled to order his priorities with no reference to the divine. Happiness, he believed, lay in the absence of fear: fear of pain, fear of poverty, fear of death. This fear was overcome through the enjoyment of the senses, yes, but this enjoyment required prudence, moderation, virtue, responsibility.

Two hundred years after Epicurus's death, his disciple Lucretius—a Roman educated in Greek philosophy, a writer of astonishing clarity—retold Epicurus's teachings in the form of a long poem that went beyond the master. *De rerum natura* (*On the Nature of the Universe* or, more literally, *On the Nature of Things*) spells out the implications of Epicurean atomism for the practice of natural science. First and foremost, Lucretius insisted, only the pure materialism of the Epicureans makes rational thought—truly *scientific* thought—possible. As long as men insist on believing in a supernatural Designer or Mover, even a benevolent one, they will suffer from "terror and darkness of the mind."

Clarity of thought, the ability to grasp physical reality as it actually is, and (above all) peace of mind come only when men admit that the universe consists of nothing but the material. Nothing else: no hell below us, nothing but sky above. Lucretius was the Richard Dawkins of the ancient world, zealous in his materialism, harsh in his criticisms of those who clung to supernatural explanations. "Nothing is ever created by divine power out of nothing," he writes in Book I of *De rerum natura*.

> The reason why all mortals are so gripped by fear is that they see all sorts of things happening on the earth and in the sky with no discernible cause, and these they attribute to the will of a god. Accordingly, when we have seen that nothing can be created out of nothing, we shall then have a clearer picture of the path ahead, the problem of how things are created and occasioned without the aid of gods.[5]

By letting go of the immortality of the soul and accepting that all ends at death, the human mind achieves freedom of thought; the fear of "eternal punishment at death" does nothing but obscure understanding and distort our reasoning.

With immortality off the table, Lucretius then reasons his way to a series of assertions about the universe. The atoms that make up everything we see are in "ceaseless motion" and vary in size and shape. The earth was not made for man; if it had been, it would be much more hospitable than it is. Rather, the earth gave birth to both animals and human beings; it "alone created the human race." The soul is a real thing, but like our bodies, it is made up of material particles, of atoms—in this case, atoms "most minute." Too tiny to comprehend, they disperse into air when the body dies, and so the soul also ceases to exist.

But the most central truth of atomism, as Lucretius explains in Book II, is that all things come to an end. All natural bodies— sun, moon, sea, our own—age and decay. None are sustained or delivered by the divine. Rather, they are struck again and again by "hostile atoms" and slowly melt away. And what is true of the physical bodies within the universe is true of the universe itself: "So likewise," he concludes, "the walls of the great world . . . shall suffer decay and fall into moldering ruins. . . . It is vain to expect that the frame of the world will last forever."[6]

Like the perishing of our own bodies, the death of the universe comes without an afterlife. No god will save our souls; no god will step in to change the course of our world.

·

Lucretius was not doing science.

He made no use of Archimedes's calculations. He had no proof of his atoms, any more than the priests he excoriated had proof of their deities. There was no way for him to test his assertions.

Yet he managed to give voice to a principle that, in the centuries to come, would become the bedrock of modern science: Explanations cannot come from outside the material world. Or, as Lucretius himself wrote,

Since there is no thing outside the whole,
    there is nothing that could enter the whole and produce a change.

All that there is is *what* there is. And in accepting the absence of what he could not believe in, Lucretius hoped to set reason free.

To read relevant excerpts from On the Nature of the Universe, *visit http://susanwisebauer.com/story-of-science.*

LUCRETIUS
On the Nature of the Universe (De rerum natura)
(ca. 60 BC)

Lucretius wrote in Latin verse, the scientific prose of the ancient world. A readable, relatively modern prose translation by Ronald E. Latham is available in print only from Penguin.

Lucretius, *On the Nature of the Universe*, trans. Ronald E. Latham, Penguin Classics, revised sub. edition (paperback, 1994, ISBN 978-0140446104).

J. S. Watson's older, more literal translation is still readable, and available as an e-book.

Titus Lucretius Carus, *On the Nature of Things*, trans. John Selby Watson, Henry G. Bohn (e-book, 1851).

To make a run at the poem in a format closer to the original, try the Oxford World's Classics edition, which retains the poetic lines of the original. The translation itself is both clear and elegant.

Lucretius, *On the Nature of the Universe*, trans. Ronald Melville, Oxford World's Classics, Oxford University Press (paperback, 2009, ISBN 978-0199555147).

The Earth-Centered Universe

The most influential science book in history

In brief, all the observed order . . . would be thrown into
utter confusion if the earth were not in the middle.
——Ptolemy, *Almagest*, ca. AD 150

By the second century AD, astronomers and mathematicians had
used Aristotelian physics, Archimedean calculations, and Lucretian
principles to construct a completely erroneous model of the universe.

This universe was spherical, and it contained five types of matter: earth, water, air, and fire, plus a fifth, mysterious substance
whose existence was deduced rather than seen—the ether, thought
to fill the celestial realm.

Careful observation and rigorous deduction had yielded an
obvious conclusion: Our planet sat at the universe's center. After
all, if you hurl a handful of dirt or toss a basinful of water into
the air, it falls down; thus, earth and water were clearly seen to be
"heavy matter," meaning that they are drawn toward the center
of the universe. The earth is made of heavy matter. But since the
earth is obviously not falling through space (this was scientific:
no one could observe this movement; therefore it did not exist), it
must already be at the universe's core.*

---

*The presence of ether was necessary because of a prior logical principle: There is
no such thing as a space where nothing exists. Every space in the universe is filled
with *something*; therefore, the something was given a name.

Fire and air do not fall. In fact, fire can even be seen to reach upward. So fire and air were classified as "light matter," which constantly moves upward, *away* from the center. The stars above the earth, along with the seven independently moving celestial bodies known as the *asteres planetai* (wandering stars) did not seem to be drawn to the center; ergo, they were made of light matter. And since light matter moves more easily and more quickly than heavy matter, it seemed clear that the light stars were moving around the heavy earth. To assume the opposite would have been entirely counterintuitive.[1]

This model was confirmed by mathematical calculations. Centuries of astronomical records—some of them inherited from the stargazing Babylonians to the east, many more recorded by Greek watchers who charted the constantly moving skies from year to year and decade to decade—yielded plenty of raw data. Second-century astronomers, using the earth-centered model, could accurately calculate the future positions of both the stars and the seven wandering planets.

The mathematics involved was both complex and ingenious. To account for the movements of the planets, Greek astronomers constructed paths for them in which each planet came to a regular stop in its orbit (a "station") and then backtracked for a predictable, calculable distance ("retrogradation").

Living in a time when we can see into the heavens, it is hard for us to enter into the mind-set of ancient astronomers, who had no closer vantage point than their earth-bound eyes. But in all likelihood, the Greeks did not intend for their models to be taken as snapshots of the universe; they did not believe that, should they be suddenly transported into the heavens, they would actually *watch* Jupiter suddenly charge backward in retrograde. The mathematical patterns were just that—not realistic portrayals of the heavens, but sets of calculations that could predict where a planet or star might be, three months, or six months, or two years hence. The mathematics was a stratagem, a way of tricking the universe into revealing part of a puzzle whose true solution lay beyond their ken.

This was called "saving the phenomena"—finding geometric patterns that matched up with observational data. These calculations, constantly adjusted, did not account for every single variation in celestial movement. But they were reliable for navigators and for timekeepers, and certainly exact enough to give astronomers confidence that—with continual small adjustments demanded by new data—they were on the right path.[2]

In the middle of the second century BC, the great stargazer Hipparchus made use of additional strategies: he charted orbits for the moon and the planets on which they performed additional small loops ("epicycles") while traveling along the larger circles ("deferents"). And he calculated the center of the deferents to be not the earth itself, but a point slightly offset from the theoretical core of the universe (the "eccentric").[3]

Using all of these tricks, astronomers were able to accurately predict the future position of any given star or wanderer. The earth-centered universe, with the planets dancing in their complicated revolutions around the core, was a model that worked. And when, around AD 150, the Greek astronomer Ptolemy took on the task of assembling all of these observations and calculations into a single manual that would account for the movements of each

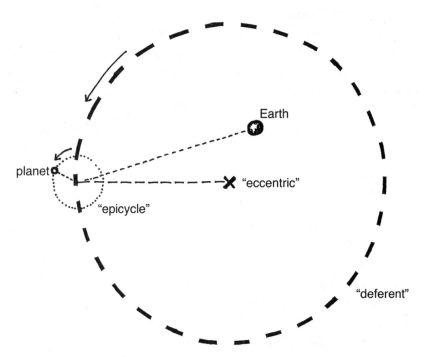

6.1 THE SCHEME OF HIPPARCHUS

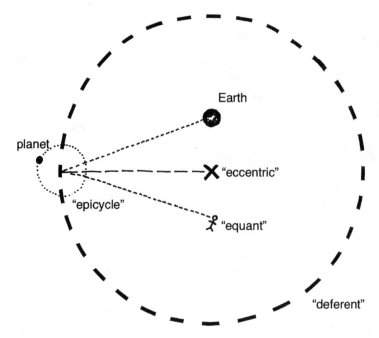

Earth

planet

"eccentric"

"epicycle"

"equant"

"deferent"

6.2 THE SCHEME OF PTOLEMY

heavenly body, Hipparchus's model was enshrined into an unquestioned system that would last for over a millennium and shape the
mind of every astronomer who gazed at the sky.

This manual, the *Almagest*, makes use not only of Hipparchus's
epicycles and eccentrics, but also of a new ploy. Ptolemy, unable to
find the exact equations that would make planets move at the same
rate all the way around their larger orbits, proposed that while the
eccentric should remain the center of the deferent, the *speed* of
planetary movement should be measured from an imaginary standing point called the *equant*.

The equant was self-defining—it was the place from which
measurement had to be made in order to make the planet's path
along the deferent proceed at a completely uniform rate. It was, in
other words, a mathematical cheat. But it was no more of a cheat
than the epicycle or the eccentric, and since it gave even more accurate predictions, it, too, became part of astronomical tradition. As
mathematician Christopher Linton points out, any planetary orbit,
no matter how complex, can be predicted by using the equant and
eccentric and by building epicycle upon epicycle—which explains

why this type of calculation remained "the cornerstone of all quantitative planetary theories" until the sixteenth century.[4]

•

For the next fourteen hundred years, the *Almagest* was almost entirely unquestioned.

In the Greek-speaking empire centered at Constantinople, the *Almagest* was continually studied and its calculations practiced. But there were no innovations, no paradigm shifts. The earth's position at the center of the universe remained a fundamental truth; Ptolemy's epicycles and equants were accepted as law.

Perhaps the very effectiveness of the *Almagest* prevented its interrogation; when the answers come out correct (or close enough; the Byzantines were content with a fairly generous margin of error), the method doesn't invite much scrutiny. Maybe, as H. Floris Cohen has argued, the familiarity of the tradition discouraged Byzantine thinkers from examining it too closely. But for whatever reason, Byzantine scholars did little with their scientific texts ("apart from bouts of intensive copying and some reshuffling," as Cohen puts it). No great questions were posed and answered by the Byzantines.[5]

Arab astronomers did little better.

Thanks to proximity, Muslim scholars had both access to the Greek texts and the language skills to make use of them. Around 820, the *Almagest* was translated into Arabic; the astronomer Ahmad al-Farghani then wrote a précis of Ptolemaic astronomy, *The Compendium of the Almagest*, which soon became the standard Arabic text on the subject. The ninth-century astronomers Thabit ibn Qurra and Muhammad ibn Jabir al-Battani, among others, proposed refinements to account for discrepancies between Ptolemy's predictions and their own observations. But the Islamic tradition, as a whole, was uninterested in scientific knowledge for its own sake. Problems of faith (the nature of the Koran and of the soul, the role of logic, the relationships of Platonism and Aristotelianism to revealed knowledge) ranked much higher in the work of Muslim astronomers. And so, like their counterparts in Constantinople, they left Ptolemy's system essentially unchallenged.[6]

To the west of the Black Sea, European scholars were even less engaged in understanding the universe.

European education, after all, was rooted in the Roman intellectual tradition. Thanks to the dominance of the Roman Empire, Roman learning had slowly supplanted Greek education. And the Roman mind-set was a practical one. It gave priority to skills (such as rhetoric) that were useful in law and politics; much less important was the investigation of the natural world, an interesting but not particularly practical pastime. New scientific pursuits withered. And as knowledge of the Greek language faded, so did awareness of the old Greek scientific texts.[7]

With the failure of the Roman political machine, the duties of education were picked up, over the course of the fifth through eighth centuries, by cathedral schools—and this learning, too, had its own biases. Bishops in the West had a vested interest in making sure there were enough educated youth to qualify as future clergymen. This required learning in the traditional liberal arts: the arts of expression (the *trivium*, made up of grammar, logic, and rhetoric) and the arts of knowledge (the *quadrivium*, encompassing arithmetic, geometry, astronomy, and music). But Christian education had inherited the Roman tendency toward pragmatism, and the *trivium* was far more useful than the *quadrivium*. A clergyman needed to be able to read, speak, and convince others. Predicting the movements of the stars, not to mention mastering the complex geometric skills needed to do so, was irrelevant.

More and more, students were given a shallow, fleeting exposure to the *quadrivium*. Rather than grappling with the difficult calculations of the *Almagest* itself, they used digests and handbooks that summarized its conclusions (spherical universe, earth at the center, rotating heavens) and left out the math. It was a physics-for-poets version of astronomy that gave them no reason to question Ptolemy's conclusions—and no reason to ask why the calculations involved were so incredibly, obfuscatingly *complicated*.

Over time, the handbooks and digests essentially replaced the *Almagest* itself. The text became rarer and rarer in the West. The educated European knew his Ptolemaic universe but knew nothing of Ptolemy. The earth-centered universe had passed into common knowledge; it was no longer a theory proposed by a single scientist that might still be disproved, but a truism authored by no one and accepted by all.

Not until the twelfth century, when the Christian kingdoms of the Spanish peninsula began to push against the Muslim dynasties to their south, did the *Almagest* itself reappear.

These Muslim dynasties had controlled the lower half of the peninsula for over four hundred years. They had carried with them their Arabic translations of Greek texts from the East; and so the libraries of southern Spain contained books that the European West had forgotten, and now had no access to. But by the 1130s, Muslim strength in the south had ebbed. The Christian king Alfonso the Battler, who had managed to draw the four kingdoms of the north together under a single joint crown, began to fight his way toward the Mediterranean, and his heirs followed his example.

By 1200, much of the south, including the prosperous city of Toledo and its extensive Arabic library, was in Christian hands.

The freedom to travel to Toledo opened up an entirely new set of texts for European scholars. And although most of them knew little Arabic and less Greek, a few—such as the prolific Gerard of Cremona, who single-handedly translated over seventy major works of science, mathematics, and astronomy into Latin—had the language skills to reintroduce these "lost" books to their colleagues.*

It took Western astronomers some time to begin making use of the new, highly technical manuals of astronomy, physics, and mathematics. Centuries of language-centered education had resulted in a Europe full of scholars who weren't practiced in the complicated geometric skills needed for a true understanding of Ptolemy. The foundation of scientific learning had well and truly decayed, and rebuilding took time.

The rebuilding accelerated dramatically when, in 1453, Constantinople fell to the Ottomans. The Greek empire of Byzantium came to a final end, and scores of Greek-speaking scholars fled away from the Turks and toward the west.

They brought some of their treasured texts with them, but primarily they brought knowledge: knowledge of the language, facility with the figures, and the conviction that the Greek intellectual

*This phenomenon is discussed in detail in Susan Wise Bauer, *The History of the Renaissance World* (W. W. Norton, 2013), Chapter 6.

legacy, little developed in times of security, was now endangered and in need of preservation.

Among them was Johannes Bessarion, a high-ranking church-man, book collector, and Aristotelian expert who had fled embat-tled Constantinople for Italy a decade before the final conquest. Now his attempts to bring Greek learning to the West gained energy. Among his other efforts, he recruited a young German professor working at the University of Vienna, Georges Peurbach, to produce a new guide to the *Almagest*: a combined translation, abridgment, and commentary.

Peurbach was himself an accomplished Ptolemaic astronomer, author of a popular student's manual that summarized the tradi-tional understanding of the universe for beginners. He knew no Greek, but he accepted Bessarion's commission and set to work on the Arabic text. He had finished the first six books when, at age thirty-eight, he grew suddenly ill. In April 1461, just before his death, he asked his student Johann Muller—a talented German mathematician, aged twenty-five—to finish the work.

Muller, better known by his Latin nickname Regiomontanus, agreed. Putting aside his own work, he spent several years fin-ishing Peurbach's project. The resulting book, the *Epitome of the Almagest*, was a readable and accurate abridgment of the *Almagest* that accepted its premises without question but did not hesitate to point out its errors (for example, that Ptolemy's system was forced to distort the size of the moon). It was the best guide yet to the complexities of the *Almagest*, but it was not widely read for another quarter century—not until 1496, when it was finally typeset and printed on one of the cutting-edge new presses that had spread across Europe.[8]

By then, Regiomontanus also was dead; he had succumbed to an obscure sickness in July of 1476, a month after his for-tieth birthday. (Translating the *Almagest* seemed to be poor for life expectancy.) But with the publication of the *Epit-ome*, both translators were hailed (posthumously) for their efforts in bringing the *Almagest* to a wider audience. "These two most celebrated men," effused the mathematician Georg Tannstetter in 1515, "magnificently restored the most noble discipline of astronomy which had almost been obliterated from human memory."[9]

Tannstetter, also a German, was speaking in part from national pride. But the *Epitome of the Almagest* turned out to be at least as influential as its source. It was soon a classroom standard, guiding young astronomers into a newly sharp understanding of the Ptolemaic universe. Through the *Epitome*, the *Almagest* regained its place as the bible of astronomy.

Within a generation, it would face a fatal challenge.

To read relevant excerpts from the Almagest, *visit http://susanwisebauer .com/story-of-science.*

PTOLEMY
Almagest
(ca. AD 150)

The *Almagest* is available in two modern translations. The R. Catesby Taliaferro translation is the first selection in Volume 16 of the "Great Books of the Western World" series, published in 1952 by Encyclopedia Britannica. It is now out of print but widely available used, as well as in most academic and many public libraries.

Robert Maynard Hutchins, ed., *Ptolemy, Copernicus, Kepler* (Great Books of the Western World, vol. 16), Encyclopedia Britannica (hardcover, 1952, ISBN 978-0852291634).

A more recent, academic but readable, and very expensive translation by G. J. Toomer was published by Princeton University Press with massive footnotes and explanatory text, making the *Almagest* easier to understand but bulking it up to nearly seven hundred pages (so probably most suitable for the true medieval-astronomy enthusiast).

Ptolemy, *Ptolemy's Almagest*, trans. G. J. Toomer, Princeton University Press (paperback, 1998, ISBN 978-0691002606).

SEVEN

The Last Ancient Astronomer

*An alternate explanation for the universe,
with better mathematics, but no more proof*

> I often considered whether there could perhaps be found a
> more reasonable arrangement of circles.
> —Nicolaus Copernicus, *Commentariolus*, 1514

Nicolaus Copernicus had a problem with the equant.

He had first encountered the *Epitome of the Almagest* in 1491, as an eighteen-year-old student at the University of Cracow. From the beginning, he had questioned those elaborate and unwieldy orbits. The entire Ptolemaic model had been built on Aristotelian physics—the properties of heavy and light elements, the tendency of the former to fall toward the center of the universe—yet it violated another one of Aristotle's central principles, which was that heavenly movements were always spherical. The equant and the epicycles, both nominally preserving the spherical orbits of the planets, actually distorted them; the only way to accept Ptolemaic orbits as spheres was to squint, really hard.[1]

And each planet required its own individual set of movements, its own particular laws. It was as if, Copernicus later wrote, an artist decided to draw the figure of a man, but gathered

> the hands, feet, head and other members for his images from
> diverse models, each part excellently drawn, but not related to a
> single body . . . the result would be monster rather than man.[2]

It was not the inaccuracies of the Ptolemaic system that bothered him; it was its inelegance.

Copernicus spent the next decade and a half studying the *Almagest* and making his own observations. Five years after his matriculation at Cracow, we find him recording the lunar eclipse of Aldebaran. Three years after that, he chronicled the conjunctions of Saturn and the moon. He lectured in Rome on mathematics, taught himself Greek, kept watching the skies.[3]

By 1514 he had formulated a more graceful theory. He wrote it out in a simple and readable form, eliminating all of the mathematics involved, and circulated it to his friends. This informal proposal, the *Commentariolus*, began with an admission that the Ptolemaic system worked reasonably well; Copernicus's primary motivation was to get rid of the mathematical gyrations that it required.

> The planetary theories of Ptolemy and most other astronomers, although consistent with the numerical data . . . present no small difficulty. For these theories were not adequate unless certain equants were also conceived. . . . I often considered whether there could perhaps be found a more reasonable arrangement of circles, from which every apparent inequality would be derived and in which everything would move uniformly about its proper center, as the rule of absolute motion requires. After I had addressed myself to this very difficult and almost insoluble problem, the suggestion at length came to me how it could be solved with fewer and much simpler constructions than were formerly used, if some assumptions (which are called axioms) were granted me.[4]

These assumptions were simple: "All the spheres revolve about the sun as their mid-point, and therefore the sun is the center of the universe." The earth was merely the center of the "lunar sphere," and it did not remain motionless. Instead, it sped in a rapid orbit around the sun (like the other spheres), moving at an amazing clip in order to complete its trip within a year, and also performed a "complete rotation on its fixed poles in a daily motion." This earthly rotation actually caused the apparent movement of the sun

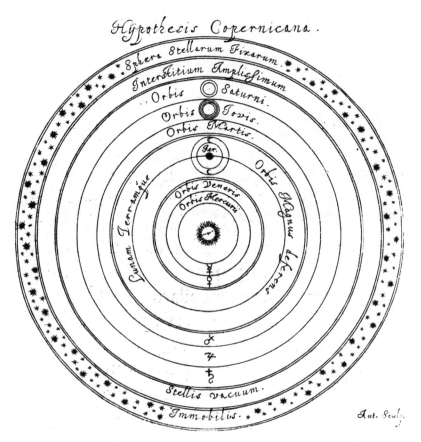

7.1 THE COPERNICAN UNIVERSE. A SEVENTEENTH-
CENTURY SKETCH BY JOHANNES HEVALIUS.

and the retrograde motions of the planets. "The motion of the
earth alone," Copernicus concluded, "suffices to explain so many
apparent inequalities in the heavens."[5]

Eighteen hundred years earlier, Aristarchus had proposed a
sun-centered universe with a moving earth; Archimedes had used
the model for his thought experiment in "The Sand-Reckoner."
The idea had never gained much traction. But Copernicus had
an advantage over previous Greek heliocentric thinkers: access to
centuries' worth of observations. The Ptolemaic system had never
worked with *complete* accuracy; there were small slippages and tiny
discrepancies in its predictions. And as more and more data were

gathered, over greater and greater spans of years, the slippages became more apparent. As Thomas Kuhn has pointed out, the movement of planets around deferents and epicycles is not unlike the movement of a clock's hands; a clock that loses a second each year will seem to be on time at the end of ten years, or even a hundred, but after a thousand years the error will become obvious.[6]

The *Almagest*, in other words, was ripe for a challenge; and Copernicus spent the next quarter century working the ideas of the *Commentariolus* up into a full-fledged manual, complete with calculations. It was a deliberate counterproposal, equal in length and complexity, and comparable in form, to the revered ancient text.

This manual, *On the Revolutions of the Heavenly Spheres*, was a masterpiece of mathematics. Like the Greek astronomers before him, Copernicus set out to "save the phenomena"—to produce a series of calculations that would line up with the data. The difference: he manipulated planets in their paths around the sun rather than the earth. And this he did successfully—without making use of Ptolemy's off-center rotations. He had, as his single private pupil Rheticus put it, "liberated" astronomy from the equant.[7]

But *On the Revolutions*, even in its final form, had massive credibility problems.

For one thing, in order to fully save the phenomena, Copernicus had to insert even more interlocking, rotating celestial spheres than his Ptolemaic colleagues had used; he had done away with the equant, but not with the overly elaborate gear system of the heavens. For another, his insistence that the earth was both rotating and hurtling through space at a ridiculously fast pace simply didn't line up with experience. It was obvious to anyone with eyes that a man who jumped up would come down in exactly the same place; the earth did not rocket out from beneath him while he was suspended in the air. Sixteenth-century physics had absolutely no way to explain this reality, and it was much more reasonable to believe that the earth was doing exactly what it appeared to be doing—standing still.[8]

On top of all this, the heliocentric theory seemed to contradict the literal interpretation of biblical passages such as Joshua 10:12–13, in which the sun and moon "stand still" rather than continuing to move around the earth. The theological problems were actually less troublesome than the scientific ones; but taken together, they cast serious doubt on Copernicus's new model.

The doubt was so pervasive, in fact, that when *On the Revolutions* was first printed in 1543, an unsigned introduction was appended to it, explaining that the heliocentric model was merely a tool for calculation, yet another mathematical trick for saving the phenomena: "For these hypotheses need not be true or even probable," the introduction assured cautious readers. "On the contrary, if they provide a calculus consistent with the observations, that alone is enough. . . . They are not put forward to convince anyone that they are true, but merely to provide a reliable basis for computation."[9]

Copernicus may not even have seen this disclaimer; it is generally thought to have been written by his friend Andrew Osiander, whom he had tasked with seeing *On the Revolutions* through the printing process. Copernicus's other writings make quite clear that he believed his model to be an accurate reflection of reality. Unlike Hipparchus, he thought that, should he be transported into the heavens, he would see the earth tracking faithfully around the sun.

Yet even in his conviction, Copernicus was aware that there was no proof for his heliocentric model. He was doing exactly the same thing that ancient astronomers had always done: constructing yet another model that gave pretty accurate answers when the calculations were run.

The preface that Copernicus himself wrote—dedicating the work to no less a personage than Pope Paul III—pointed out that the geocentric theory was itself *incertitudo mathematicarum traditionum*, an uncertain mathematical tradition, and that his model was an alternative mathematical proposal: "Because I knew that others before me had been granted the liberty of constructing whatever circles they pleased in order to demonstrate astral phenomena," Copernicus explains, "I thought that I too would be readily permitted to test . . . demonstrations less shaky than those of my predecessors."[10]

Demonstrations less shaky: Copernicus was consciously refusing to assert that his heliocentrism *was* reality. Like the Greek atomists, he had actually stumbled onto the truth. And, like the atomists, he had no method that would confirm his conclusions. He knew this perfectly well.

The rhetorical distancing of himself from his own beliefs worked, for a time. Pope Paul III happily accepted the dedication, and *On the Revolutions* remained on the church's good side for the next seventy years. But within the work itself, Copernicus's real beliefs occasionally pop to the surface with suppressed, irresistible energy. Halfway through the first chapter, he lays down the most controversial assertion of his system: that only the mobility of the earth and the sun's position at "the centre of the world" explain the motions of the stars. "The harmony of the whole world teaches us their truth," he writes, "if only—as they say—we would look at the thing with both eyes."[11]

But human eyes, even both of them, could not prove the heliocentric system.

To read relevant excerpts from the Commentariolus *and* On the Revolutions of the Heavenly Spheres, *visit http://susanwisebauer.com/story-of-science.*

If you, like Copernicus's friends, prefer the conclusions without the math, read the *Commentariolus*, which is considerably shorter than the full text of *On the Revolutions*. If you enjoy wrestling with geometric proofs, choose the longer work.

NICOLAUS COPERNICUS
Commentariolus
(1514)

The *Commentariolus* is included, along with a summary of Copernicus's work written by his champion Rheticus (the *Narratio prima*) and a letter written by Copernicus disproving the calculations of the astronomer Johannes Werner (the *Letter against Werner*), in the paperback *Three Copernican Treatises*.

Edward Rosen, trans., *Three Copernican Treatises*, 2nd revised edition, Dover Publications (paperback, 2004, ISBN 978-0486436050).

NICOLAUS COPERNICUS
On the Revolutions of the Heavenly Spheres
(1543)

The early-twentieth-century translation by Charles Glenn Wallis
has been reprinted in paperback.

Nicolaus Copernicus, *On the Revolutions of the Heavenly Spheres*,
   trans. Charles Glenn Wallis, Prometheus Books (paperback,
   1995, ISBN 978-1573920353).

Another edition of the same translation, with introductory essay
and notes by Stephen Hawking, can be found in

Nicolaus Copernicus, *On the Revolutions of the Heavenly Spheres*,
   trans. Charles Glenn Wallis, ed. Stephen Hawking, Running
   Press (paperback, 2002, ISBN 0-7624-2021-9).

THE BIRTH
OF THE METHOD

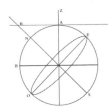

Francis Bacon, *Novum organum* (1620)

William Harvey, *De motu cordis* (1628)

Galileo Galilei, *Dialogue concerning the Two Chief
World Systems* (1632)

Robert Boyle, *The Sceptical Chymist* (1661)

Robert Hooke, *Micrographia* (1665)

Isaac Newton, *Philosophiae naturalis principia mathematica*
(1687/1713/1726)

EIGHT

A New Proposal

*A challenge to Aristotle,
and the earliest articulation of the scientific method*

The logic now in use . . . does more harm than good.
—Francis Bacon, *Novum organum*, 1620

Copernicus's scheme was a century old, but nothing had yet changed. His theory remained an outlier. It made no sense, in terms of Aristotelian physics; he hadn't managed to explain why, if the earth was circling the sun, its motion through the air was imperceptible to people on the earth's surface.

Thirty years after the publication of *On the Revolutions*, a possible solution was offered by the Danish astronomer Tycho Brahe, a fan of the Copernican system ("This innovation expertly and completely circumvents all that is superfluous or discordant in the system of Ptolemy," he wrote). The heliocentric system demanded that the earth, "that hulking lazy body, unfit for motion," move as quickly "as the aethereal [stars]." What if, instead, the earth remained still with the sun, moon, and stars rotating around it—while the other five planets orbited the sun?[1]

Tycho is now best known as the first discoverer of a new star, which he called *nova stella*—a term that survives into modern times as *supernova*. His ingenious combination system solved the physics problem and explained the motions of the skies. But like Copernicus, Tycho could offer no *proof.* Without evidence, neither his system nor Copernicus's was more compelling than the Ptolemaic explanation.

There was still no real reason to rethink Ptolemy, or to challenge Aristotle's authority.

•

In 1603, Francis Bacon, London born, was forty-three years old: a trained lawyer and amateur philosopher, happily married, politically ambitious, perpetually in debt.

He had served Elizabeth I of England loyally at court, without a great deal of recognition in return. But now Elizabeth was dead at the age of sixty-nine, and her crown would go to her first cousin twice removed: James VI of Scotland, James I of England.

Francis Bacon hoped for better things from the new king, but at the moment he had no particular "in" at the English court. Forced to be patient, he began working on a philosophical project he'd had in mind for some years—a study of human knowledge that he intended to call *Of the Proficience and Advancement of Learning, Divine and Human.*

Like most of Bacon's undertakings, the project was ridiculously ambitious. He set out to classify all learning into the proper branches and lay out all of the possible impediments to understanding. Part I condemned what he called the three "distempers" of learning, which included "vain imaginations," pursuits such as astrology and alchemy that had no basis in actual fact; Part II divided all knowledge into three branches and suggested that natural philosophy should occupy the prime spot. Science, the project of understanding the universe, was the most important pursuit man could undertake. The study of history ("everything that has happened") and poesy (imaginative writings) took definite second and third places.[2]

For a time, Bacon didn't expand on these ideas. The *Advancement of Learning* opened with a fulsome dedication to James I ("I have been touched—yea, and possessed—with an extreme wonder at those your virtues and faculties . . . the largeness of your capacity, the faithfulness of your memory, the swiftness of your apprehension, the penetration of your judgment, and the facility and order of your elocution. . . . There hath not been since Christ's time any king or temporal monarch which hath been so learned in all literature and erudition, divine and human"), and this groveling soon yielded fruit. In 1607 Bacon was appointed as solicitor general, a position he had coveted for years, and over the next decade or so he poured his energies into his government responsibilities.

He did not return to natural philosophy until after his appointment to the even higher post of chancellor in 1618. Now that he had battled his way to the top of the political dirt pile, he announced his intentions to write a work with even greater scope—a new, complete system of philosophy that would shape the minds of men and guide them into new truths. He called this masterwork the *Great Instauration*: the Great Establishment, a whole new way of thinking, laid out in six parts.

Part I, a survey of the existing "ancient arts" of the mind, repeated the arguments of the *Advancement of Learning*. But Part II, published in 1620 as a stand-alone work, was something entirely different. It was a wholesale challenge to Aristotelian methods, a brand-new "doctrine of a more perfect use of reason."[3]

Aristotelian thinking relies, heavily, on deductive reasoning—for ancient logicians and philosophers, the highest and best road to the truth. Deductive reasoning moves from general statements (premises) to specific conclusions.

MAJOR PREMISE: All heavy matter falls toward the center of the universe.

MINOR PREMISE: The earth is made of heavy matter.

MINOR PREMISE: The earth is not falling.

CONCLUSION: The earth must already be at the center of the universe.

But Bacon had come to believe that deductive reasoning was a dead end that distorted evidence: "Having first determined the question according to his will," he objected, "man then resorts to experience, and bending her to conformity with his placets [expressions of assent], leads her about like a captive in a procession." Instead, he argued, the careful thinker must reason the other way around: starting from specifics and building toward general conclusions, beginning with particular pieces of evidence and working, inductively, toward broader assertions.[4]

This new way of thinking—*inductive reasoning*—had three steps to it. The "true method," Bacon explained,

> first lights the candle, and then by means of the candle shows the way; commencing as it does with experience duly ordered and digested, not bungling or erratic, and from it deducing axioms, and from established axioms again new experiments.

In other words, the natural philosopher must first come up with an idea about how the world works: "lighting the candle." Second, he must test the idea against physical reality, against "experience duly ordered"—both observations of the world around him and carefully designed experiments. Only then, as a last step, should he "deduce axioms," coming up with a theory that could be claimed to carry truth.[5]

Hypothesis, experiment, conclusion: Bacon had just traced the outlines of the scientific method.

It was not, of course, fully developed. But Part II of Bacon's *Great Instauration* was a clear challenge to the deductive thinking of the Aristotelian corpus. Bacon even named it *Novum organum* ("New Tools"), after Aristotle's logical treatises titled *Organon*. On the cover of the *Novum organum*, Bacon placed a ship—his new inductive method—sailing triumphantly past the Pillars of Hercules, the mythological pillars that marked the farthest reach of Hercules's journey to the "far west." Identified by most ancient authors as the promontories on either side of the Strait of Gibraltar, the Pillars represented the outermost boundaries of the ancient world, the greatest extent of the old way of knowledge.[6]

Barely a year after publication of the *Novum organum*, Francis Bacon was accused by his enemies at court of taking bribes. And although he protested that he had "clean hands and a clean heart," he was unable to disprove the charges. He was removed from the chancellorship, ordered to pay a fine, and briefly imprisoned in the

FRANCISCI

DE VERULAMIO,

Summi Angliæ

CANCELARII,

Instauratio

magna.

Multi pertransibunt & augebitur scientia.

Sim: Pass: sculp:

Anno

LONDINI
Apud Joannem Billium
Typographum
Regium.

1620.

8.1 THE NOVUM ORGANUM

Tower of London; although James I ultimately rescinded his fine
and pardoned him, the wind had been thoroughly taken out of his
sails. He died five years later of pneumonia, without coming close
to finishing his *Great Instauration*.[7]

But the *Novum organum* continued to shape the seventeenth-
century practice of science. In 1662, King Charles II granted a
royal charter to the Royal Society of London for Improving Natural Knowledge, a gathering of natural philosophers who were
committed to the experimental method of science; they were all
students of the *Novum organum*, devotees of the Baconian methods. The poet Abraham Cowley, himself an enthusiastic amateur
scientist, wrote the Royal Society's dedicatory epistle; it was all in
praise of Francis Bacon, who had overthrown ancient authority
with "true reason":

> Authority, which did a body boast,
> Though 'twas but air condens'd, and stalk'd about,
> Like some old giant's more gigantic ghost,
> To terrify the learned rout
> With the plain magic of true reason's light,
> He chas'd out of our sight,
> Nor suffer'd living men to be misled
> By the vain shadows of the dead.

The experimental method finally allowed man to look directly at
nature, rather than wrestling with logic:

> From words, which are but pictures of the thought,
> Though we our thoughts from them perversely drew
> To things, the mind's right object, he it brought . . .
> Who to the life an exact piece would make,
> Must not from other's work a copy take . . .
> No, he before his sight must place
> The natural and living face;
> The real object must command
> Each judgment of his eye, and motion of his hand.[8]

The lines are, self-consciously, modeled on Lucretius's praise of Epicurus. Just as Epicurus had broken the bonds of superstition with his atomism, so Bacon broke the bonds of Aristotle with the experimental method.

"Bacon's grand distinction," observed Macvey Napier in a classic 1818 speech, "lies in this, that he was the first who clearly and fully pointed out the rules and safeguards of right reasoning in physical inquiries." Finally, a method was in place that would allow natural philosophers to "look with both eyes," as Copernicus had asked, and come to conclusions based on their observations.[9]

To read relevant excerpts from the Novum organum, *visit http://susan wisebauer.com/story-of-science.*

FRANCIS BACON
Novum organum
(1620)

Book I begins with "Aphorisms," brief independent statements that lay out Bacon's objections to the current methods in use in natural science; Book II develops his alternative proposal.

The nineteenth-century translation by James Spedding and Robert Ellis is still the most commonly reprinted. It can be read in multiple free e-book versions, such as

The Philosophical Works of Francis Bacon, trans. and ed. James Spedding and Robert Ellis, vol. 4, Longman (e-book, 1861, no ISBN).

A more recent translation with introduction, outline, and explanatory notes is

Francis Bacon, *The New Organon,* ed. Lisa Jardine and Michael Silverthorn, Cambridge University Press (paperback and e-book, 2000, ISBN 978-0521564830).

The notes are useful, but the translation, while more contemporary, is not always clearer. For example, the Spedding and Ellis translation of Aphorism XII in Book I reads:

> The logic now in use serves rather to fix and give stability to the errors which have their foundation in commonly received notions than to help the search after truth. So it does more harm than good.

Compare the Jardine and Silverthorn translation:

> Current logic is good for establishing and fixing errors (which are themselves based on common notions) rather than for inquiring into truth; hence it is not useful, it is positively harmful.

NINE

Demonstration

*The refutation of one of the greatest ancient authorities
through observation and experimentation*

Therefore it will be profitable . . . by the frequent dissection
of living things, and by much ocular testimony, to discern
and search the truth.
                    —William Harvey, *De motu cordis*, 1628

Francis Bacon, still untarred by scandal, was at the height of his
political career when William Harvey carried out his first public
dissections in London.

Harvey was thirty-seven years old, short, energetic, a machine-
gun lecturer who mixed English and Latin as he explained the
structure of the corpse in front of him, pointing out the internal
organs with the "fine white rods" provided for him by the Royal
College of Surgeons. He had just been appointed Lumleian Lecturer in Anatomy, a post that required him to teach a twice-weekly
class on the human body and, during the winter (when the corpse
would putrefy more slowly), to supplement his lectures with the
dissection of an executed felon.[1]

He was expected to base his work on the classic anatomy texts
of Galen, the second-century physician who had supplemented his
traditional Hippocratic education with years of animal dissection.
Galen had argued that an understanding of the body could come
only through knowledge of its structures; his numerous writings
and anatomical studies, reintroduced to European physicians by

twelfth-century translators, had become the foundation of all contemporary medical knowledge.

But Galen, a Greek-speaking citizen of the Roman Empire, had inherited a double taboo against human dissection. The ancient Greek belief that proper burial was necessary before the soul could enter the Elysian Fields had given way to the Roman superstition that the unburied roamed the earth in misery; and even those rationalists who believed in neither the soul nor the afterlife generally accepted that human bodies should be decently interred, not cut apart. Aside from a small pocket of rogue anatomists in Alexandria (Herophilus and Erasistratus, working in the third century BC), Greek and Roman medical men studied dogs, cats, bulls, and occasionally apes, extrapolating their discoveries to the still-mysterious human body.[2]

With the resurrection of the Galenic tradition came a growing impulse to check Galen's findings against actual human anatomy; and the Christian West was, on the whole, less superstitious about the body than the ancients had been. Human dissection seems to have been reintroduced into university lectures, slowly and sporadically, beginning around 1315 or so, and the practice gained traction when Pope Sixtus IV declared, in 1482, that there was no theological reason to avoid human dissection as long as the remains were given Christian burial afterward.[3]

In Harvey's day, medical training at Italian universities was the most likely to make use of human subjects for anatomy; Jacopo Berengario da Carpi, a lecturer at Bologna, was said to have dissected "several hundred" human bodies, and his illustrated guide to the human body is one of the earliest works of anatomy to contain accurate drawings of the structure of the heart, appendix, and uterus. And it was a professor at the University of Padua, Andreas Vesalius, who had mounted the most serious challenge yet to Galen's authority: *De humani corporis fabrica*, a massive manual of anatomy, illustrated with scores and scores of detailed drawings done from actual dissections. Accurate anatomical knowledge, argued Vesalius, had to be the foundation of all medical learning and practice, and anatomists had to look with their own eyes; they should never simply rely on the teachings of past authorities, as was too often done in university training. "As things are now taught in the schools," he complained, "with days wasted on ridiculous questions, there is very little offered . . . that could not better be taught by a butcher in his shop."[4]

The *Fabrica*, published in 1543, established a particularly strong practice of hands-on anatomy at Padua; and William Harvey had chosen to travel to Padua for his medical degree. There he was taught in the tradition of Vesalius, who exhorted his students to work hard at "dissecting and examining in person the fabric of the human body, and then carefully comparing it with the teachings of Galen."[5] This close observation—the "looking with both eyes" that Copernicus had recommended, in frustration—had revealed to Vesalius plenty of places where Galen's conclusions were obviously, demonstrably, clearly *wrong*.

Yet even half a century after the *Fabrica*, Harvey was expected to base his lectures on Galen, the "father of medicine"; those texts were treated with as much respect and deference as Aristotle's *Physics* and Ptolemy's *Almagest*.

•

In his very first year as Lumleian Lecturer, Harvey flat-out contradicted the master.

"It is shown by the application of a ligature," his lecture notes read, "that the passage of the blood is from the arteries into the veins. Whence it follows that the movement of the blood is constantly in a circle, and is brought about by the beat of the heart."[6]

This was not at all what Galen had taught.

Galen's own observations, mostly of animal anatomy, had revealed that the heart had two chambers, and that the blood in the right chamber was darker than the blood in the left. But dissection couldn't reveal exactly how these chambers *worked*. So Galen theorized that each chamber pumped a different kind of blood.

The darker blood in the right-hand chamber, he believed, was actually manufactured by the liver. The stomach digested food into a milky nourishing liquid called *chyme*, which was then sent to the liver, which transformed the chyme into "venous blood" and sent it out through veins (which, Galen explained, all *originated* in the liver) to nourish the organs of the body. Some of it went into the right chamber of the heart, where it was then sent to feed the lungs.

Arteries, on the other hand, carried a different sort of blood. Air—*pneuma*, "vital spirits"—entered the body through the lungs and then traveled from the lungs to the left-hand chamber of the heart, where it was combined with venous blood and transformed into "arterial blood"—thinner, brighter, and quicker than venous blood. Then the arteries carried these vital spirits, by way of the arterial blood, to the organs as well. Food and vital spirits, venous blood and arterial blood—the organs of the body needed both to function.

But neither kind of blood needed a pump to send it along.

The organs were thought to attract venous blood to themselves whenever they were in need of food, with a sort of sucking power. Arterial blood was moved along by the pulsating motions of the arteries themselves. The function of the heart was merely to suck in air and venous blood, and transform them into arterial blood.

So, the pulsing action of the heart (the Greeks could hear a heartbeat as well as anyone else) was explained away as an *effect*, not a cause. Draw in a breath, and the pneuma rushes into your lungs and then travels from the lungs to the left chamber of the heart; when it comes into contact with the venous blood there, it heats the venous blood to the boiling point. As the heated blood expands, the heart expands to hold it ("diastolic" action, from the Greek *diastole*, "dilation"). Then the blood cools and rushes out into the arteries; the heart relaxes and contracts back to its original size ("systolic" action, from *systellein*, "to contract").

There was one problem with this scheme; it required venous blood to move somehow from the right-hand chamber of the heart over to the left-hand chamber, so that it could be combined there with pneuma. And there didn't seem to be any connection between the two chambers.

So Galen acted against his own observations, allowing (in an Aristotelian sort of way) his overall theory to dictate physical reality. There must be pores, he wrote, too small for the eye to see, between the right and left chambers of the heart; through these pores, venous blood seeped from the right to the left.[7]

This was exactly the sort of deductive thinking that Francis Bacon had deplored.

Harvey had multiple problems with the Galenic theory. First, he could find no observational proof of these infinitesimal pores. Second, so far as he could see, the two chambers of the heart were very much the same in structure; how could this be, if they performed such different functions? Third, arterial blood and venous blood seemed to be exactly the same in every quality except their color, so it seemed unlikely that one contained digested food and the other air. Fourth, the artery that led out of the left chamber of the heart and the vein that led out of the right seemed to have about the same capacity—but according to Galen, the artery carried arterial blood to the entire body, and the vein only carried venous blood to the lungs. Why did the lungs need so much blood?[8]

Harvey needed a new theory that was based on his structural observations and that he could arrive at inductively—by examining the heart and its function, and by testing both his theories and Galen's by experiment. For a decade and half, he worked to elaborate this theory. He measured the amount of blood pumped out by the heart with each beat, and realized that Galen's theory required more blood than the body could hold. He dissected the hearts of animals and saw that they behaved like muscles, and that the contraction of the heart was the active, working phase of its movement—not a relaxing, as Galen had theorized. He painstakingly dissected veins and arteries in search of the valves that helped blood to flow in one direction—not wash back and forth, as Galen's model demanded. He also dissected animals that were still alive, a practice revolting to us now but legitimate medical research back then, so that he could watch the heart in action. "No one," he told a visitor later in his life, "has ever rightly ascertained the use or function of a part who has not examined its structure, situation, connections by means of vessels and other accidents in various animals, and carefully weighed and considered all he has seen."[9]

All of this was Baconian, the new experimental method in practice—although both Francis Bacon and William Harvey would have rejected this idea with scorn. Far from criticizing Aristotle, Harvey held him up as an authority, primarily because Aristotle had believed that the heart was the most important organ in the body (Galen gave this honor to the liver). Harvey had actually treated Bacon for gout during Bacon's years as lord chancellor, and hadn't taken to him; he remarked that Bacon wrote philosophy like a politician (an insult), and wrote to a friend that the chancellor's eyes were "like the eye of a viper."[10]

For his part, Bacon had a low opinion of doctors. An entire section of the *Advancement of Learning* is devoted to the shortcomings of the field of medicine. In Bacon's view, doctors as a race were far too lax about rigorously searching out causes and effects, too willing to try superstitious or irrational cures, careless about the makeup of their medicines, and insensitive to their patients' pain.

But Harvey kept on with his Baconian experimentations, building up layer after layer of support for his new theory. In 1628 he published his findings in Latin as *De motu cordis* ("On the Motion of the Heart"); an English translation was in circulation by 1653.

In *De motu cordis*, Harvey proposed that blood was pumped from the right side of the heart into the lungs and then moved from the lungs to the left side of the heart, and from there throughout the body by way of the arteries. It then returned to the heart by way of the veins, creating a complete circle. Every part of this system was demonstrable except for one—the pathway by which the blood moved from the arteries back into the veins.

Ironically, in this one area Harvey was forced to use the same kind of reasoning he had deplored in Galen. He suggested that there must be tiny, threadlike connections between the veins and the arteries, too small for his eyes to see.

This turned out to be the case; the capillaries were finally viewed, with the help of new microscopic technology, by the Italian physician Marcello Malpighi in 1661, just four years after Harvey's death. But even in the absence of this one particular proof, the theory of the motion of the heart and the circulation of the blood lined up, in every other way, with the visual evidence provided by dissection.

To the end of his life, Harvey kept demonstrating the truth of his system, and kept experimenting to make absolutely sure that his criticisms of Galen were just. When he was in his seventies, he was still piling up the evidence. In March of 1651 he wrote to a friend that he had "lately tried" a new experiment in which he had tied off the right chamber of the heart of "a man who had been hanged" and forcibly filled it with water. The result was simple: "not a drop of water or of blood made its escape" into the left ventricle. Galen's pores did not exist.[11]

"You may try this experiment as often as you please," Harvey's account of this experiment concludes. "The result you will still find to be as I have stated it." Replicable, experimental proof had been used to create an entirely new theory of the body; the ancient authority of Galen had been dethroned.

To read relevant excerpts from De motu cordis, *visit http://susanwise bauer.com/story-of-science.*

WILLIAM HARVEY
De motu cordis
(1628)

The original English translation of 1653 has been republished as a Dover paperback, along with the accompanying essay *De circulatione sanguinis*. The language is slightly archaic but still readable.

William Harvey, *The Anatomical Exercise*, ed. Geoffrey Keynes, Dover Publications (paperback, 1995, ISBN 978-0486688275).

The widely read nineteenth-century translation by Robert Willis is available as a free e-book, as "An Anatomical Disquisition on the Motion of the Heart and Blood in Animals."

Robert Willis, trans., *The Works of William Harvey, M.D.*, Sydenham Society (e-book, 1847, no ISBN).

It has also been reprinted in paperback by Prometheus Books.

William Harvey, *On the Motion of the Heart and Blood in Animals*, trans. Robert Willis, Prometheus Books (paperback, 1993, ISBN 978-0879758547).

TEN

The Death of Aristotle

The overthrow of ancient authority
in favor of observations and proofs

We do have in our age new events and observations
such that, if Aristotle were now alive, I have no doubt he
would change his opinion.
                    —Galileo Galilei, *Dialogue concerning*
                    the Two Chief World Systems, 1632

When twenty-one-year-old William Harvey arrived in Padua
to study at the School of Medicine, Galileo Galilei was already
ensconced there as professor of mathematics.

Galileo, fourteen years Harvey's senior, was a wildly popular
lecturer; he was forced to use the university's largest hall, seat-
ing two thousand, to accommodate the students who came from
all over the continent to hear him. But we have no record that
Harvey attended. "Mathematics" was a wide term, and Galileo's
lectures were expected to encompass arithmetic, geometry, astronomy, physics, and the arts of military fortification and engineer-
ing. These were subjects that did not necessarily cross Harvey's
curriculum; medical education, then as now, tended to be both
intense and narrow.[1]

Galileo had come to Padua after three unhappy years at the
University of Pisa. He had become increasingly critical of Aristotelian physics, and his opinions had not endeared him to the tradi-
tionalists at Pisa. When he tactlessly evaluated a dredging machine
designed by a university patron as "useless" (and demonstrated just

how badly the proposed model would perform), the embarrassed inventor joined the ranks of his enemies.[2]

So Galileo was quick to accept the post at Padua when it was offered to him in 1592. But Pisa had not been a waste of time. While there, he had written and circulated an unpublished set of essays on force and motion, De motu, which made a first step toward solving the knottiest problems of the Copernican system.

Galileo himself was not, at this point, particularly focused on the skies. His concerns lay closer to the earth. Later in his life, he wrote that he had begun to doubt Aristotelian physics when he saw large hailstones and small ones hitting the ground side by side. According to Aristotle, this could happen only if all of the large hailstones fell from higher up, because large objects fall faster than small ones. (Objects that fall do so because they seek their "natural place"—for objects made of "heavy" elements, the center of the universe. Large objects, because they contain a greater concentration of heavy elements, fall faster than small objects.)[3]

Galileo could not believe that *all* of the large hailstones were falling from farther up in the sky. And the essays in De motu show that, while at Pisa, he had carried out a series of experiments and demonstrations that clearly contradicted Aristotle's physics of motion.

Nearly seventy years later, the first biography of Galileo—written by the Italian mathematician Vincenzo Viviani, who had served as the old man's student assistant—insisted that Galileo had disproved Aristotle by repeatedly dropping unequal weights "from the height of the Leaning Tower of Pisa" and watching them hit the ground simultaneously. Viviani's biography is filled with errors of time and place, casting considerable doubt on this dramatic account. But, as Stillman Drake points out in his classic study of Galileo's thought, it is quite likely that Galileo carried out public demonstrations so that his Pisan students could see the proofs of his studies with their own eyes. He believed, firmly, that the truth could always be demonstrated. "Truth . . . is not so deeply concealed as many have thought," he wrote in Chapter 9 of De motu. "[It] is shown to us by Nature so openly and clearly that nothing could be plainer or more obvious."[4]

10.1 GALILEO'S EXPERIMENT

Yet Galileo lived in a world where this Baconian method—the demonstration of truth through repeated experimentation—was still junior to received authority, still secondary to tradition. Forty years later, Galileo would write scathingly of a Venetian philosopher who attended a public dissection, carried out by a celebrated anatomist who intended to disprove Aristotle's insistence that all nerves originated in the heart:

> The anatomist showed that the great trunk of nerves, leaving the brain and passing through the nape, extended on down the spine and then branched out through the whole body, and that only a single strand as fine as a thread arrived at the heart. Turning to [the philosopher], on whose account he had been exhibiting and demonstrating everything with unusual care, he asked this man whether he was at last satisfied and convinced that the nerves originated in the brain. . . . The philosopher, after considering for awhile, answered: "You have made me see this matter so plainly and palpably that if Aristotle's text were not contrary to it, stating

clearly that the nerves originate in the heart, I should be forced to admit it to be true."[5]

Seeing was not yet believing.

Galileo did not publish *De motu*, most likely because he had not found satisfactory answers to some of his most central questions. But he continued to *look*. Some fifteen years after his arrival in Padua, he learned of a new tool that could extend the range of his eyes; the spectacle maker Hans Lippershey, practicing his craft in the Netherlands, had put together the convex lenses he used to correct farsightedness and the concave lenses that aided the nearsighted, creating a new instrument. On October 2, 1608, Lippershey asked the Dutch legislature, the States General, to grant him patent protection for this "telescope."[6]

The States General bought a telescope from Lippershey but refused to give him a patent, and within a year telescopes were being assembled all over Europe. Galileo seems to have first encountered a telescope while visiting Venice, sometime in 1609, and upon returning home he immediately set to work grinding his own lenses and improving the instrument's refraction.

Lippershey's instrument was only slightly more useful than the naked eye; Galileo managed to refine the magnification to about 20X. Barely a year after his first glimpse through a telescope, he published a study of the skies based on his observations. "The Sidereal ['starry'] Messenger, unfolding great and marvellous sights," the frontispiece read, "respecting the Moon's Surface, an innumerable number of Fixed Stars, the Milky Way, and Nebulous Stars, but especially respecting Four Planets which revolve round the Planet Jupiter at different distances and in different periodic times, with amazing velocity."[7]

Through his telescope, Galileo had seen mountains and valleys on the moon, many more stars than could be seen with the eye alone, and nebulae—cloudy heavenly bodies made up of clusters of individual stars. But the four planets orbiting Jupiter were, in his words, "the matter . . . most important in this work." They had never before been seen; at first, Galileo had thought them to be

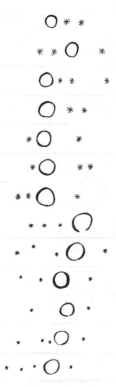

10.2 GALILEO AND JUPITER. A REPRODUCTION
OF THE SKETCH GALILEO MADE OF HIS
TELESCOPIC OBSERVATIONS OF JUPITER.

newly visible fixed stars, but when he looked at them again on the following day, they had *moved*.

And they kept moving, in and out of sight, to the left and to the right of Jupiter itself. Over the course of a week, Galileo was able to sketch out their progression and come to an inevitable conclusion: "They perform their revolutions about this planet . . . in unequal circles." Galileo's observations provided unequivocal proof that not all heavenly bodies revolve around the earth. And in the months after publication of *The Sidereal Messenger*, Galileo used his telescope to observe the changing phases of Venus—inexplicable in the Ptolemaic system, making sense only if Venus were, in fact, traveling around the sun.[8]

Aristotelian physics, already dealt a mortal blow by Galileo's experiments with weights, had been administered a coup de grâce. But the corpse still had its loyalists. The chief philosopher at Padua, an Aristotelian named Cesar Cremonini, simply refused to look through Galileo's telescope: "What do you say of the leading philosophers here," Galileo wrote bitterly to the astronomer Johannes Kepler, "to whom I have offered a thousand times to show my studies, but who, with the lazy obstinacy of a serpent who has eaten his fill, have never consented to look at the planets, or moon? . . . To such people . . . truth is to be sought, not in the universe or in nature, but (I use their own words) by comparing texts!"[9]

An epic battle was shaping up: between ancient authority and present observation, Aristotelian thought and Baconian method, text and eye. Galileo himself had no quarrel with Aristotelian logic or philosophy, but he refused to take the great man's physics for granted. Later, in *Dialogue concerning the Two Chief World Systems*, he put the objections of his opponents in the mouth of his tradition-bound character Simplicius: "If Aristotle is to be abandoned, whom should we have for a guide?" asks Simplicius. "We need guides in forests and in unknown lands," responds Galileo's spokesman in the dialogue, the scholar Salviati, "but on plains and in open places only the blind need guides." Philosophy might still be an unknown, thickly wooded place, but in Galileo's opinion, physics and astronomy were now clear spaces where anyone with wits and eyes could see the truth; the earth had been "lifted up out of darkness and exposed to the open sky." Aristotle himself, he argued, would have appreciated the new discoveries—and would have been willing to adjust his physics accordingly.[10]

But Aristotle's adherents did not agree. And they included not just philosophers and scholars (who had no power to do anything but dissent), but also the churchmen who were in charge of the Inquisition, which had a great deal of power indeed.

By the beginning of the seventeenth century, the theology of the church centered at Rome had been thoroughly Aristotelianized. Thomas Aquinas's great thirteenth-century synthesis of Christian revelation and Aristotelian metaphysics had infused Roman Christianity. A central element of this synthesis was a careful separation between those things that could be discovered through the exercise of human reason and the use of the senses (such as truths about the natural world) and those realities that could be understood only through divine revelation (such as the nature of God). While this separation might seem to line up nicely with Galileo's own point of view, it actually introduced a fatal contradiction: the Bible was God's revelation of himself and so fell into the second category of truth—that which *cannot* be understood through the senses or through reason. It was a text that must be accepted, not analyzed— much like Aristotle's own writings.*

Galileo's discoveries were doubly troubling: they contradicted both Aristotle *and* the literal meaning of several biblical passages. And, thanks to the telescope, the movements of the planets could no longer be explained away as a mathematical trick to "save the phenomena."

In 1615, Pope Paul V ordered the cardinal Robert Bellarmine to begin a formal investigation of Galileo's work and its implications. Galileo himself had not yet written anything that explicitly argued the heliocentric position, although his observations in *The Sidereal Messenger* certainly implied that he accepted it. So, after a year's sleuthing, Bellarmine recommended that not Galileo's work, but Copernicus's *On the Revolutions*, be placed on the list of heretical, condemned texts (the *Index Librorum Prohibitorum*). He also warned Galileo, in a private but official meeting, to abandon public agreement with Copernicus.

In a letter to the Carmelite monk Paolo Antonio Foscarini, who had argued that Copernicus's system didn't contradict scripture at all, Bellarmine suggested that the heliocentric model remain a mathematical one alone. "It seems to me," Bellarmine wrote,

> that [you] and Mr. Galileo are proceeding prudently by limiting yourselves to speaking suppositionally and not absolutely. . . . For there is no danger in saying that, by assuming the earth moves and the sun stands still, one saves all the appearances better than by

---

*A more detailed analysis of the relationship between Aristotelian philosophy and Christian theology is beyond the scope of this book. William C. Placher provides a useful summary in "The Fragile Synthesis," Chapter 10 of *A History of Christian Theology: An Introduction* (John Knox Press, 1983).

postulating eccentrics and epicycles. . . . However, it is different to want to affirm that in reality the sun is at the center of the world . . . and the earth . . . revolves with great speed around the sun; this is a very dangerous thing, likely not only to irritate all scholastic philosophers and theologians, but also to harm the Holy Faith by rendering Holy Scripture false.[11]

"Speaking suppositionally" would protect both the Bible and Aristotle: a dual victory.

Bellarmine was not suggesting that telescopic evidence be ignored. Rather, he lacked the mathematics to follow Galileo's conclusions. As far as he was concerned, heliocentrism had no proof to support it (and, to be fair, Galileo had not yet solved the problem of the earth's apparent immobility). In fact, Bellarmine was willing to reexamine the matter, should proof be found:

If there were a true demonstration that the sun is at the center of the world . . . and that the sun does not circle the earth, but the earth circles the sun, then one would have to proceed with great care in explaining the Scriptures that appear contrary, and say rather that we do not understand them than that what is demonstrated is false. But I will not believe that there is such a demonstration, until it is shown me.[12]

This declaration rang, in Galileo's ears, as a challenge. In 1616 he began to circulate a manuscript essay called "On the Tides," which argued that the movements of the seas could be explained only if the earth were both rotating on its axis *and* orbiting the sun. His regular correspondent Johannes Kepler—now holding the position of imperial mathematician to the Holy Roman emperor—was simultaneously working out new orbits for the planets that gave the Copernican model even greater accuracy.

Over the next sixteen years, Galileo tackled the remaining problems of the heliocentric model, one at a time. He came to a satisfactory explanation for the earth's apparent immobility, using as an analogy the dropping of an object from a ship's mast: even if the ship moves, the object falls to the base of the mast, every time.

He worked on the continuing problems of the tides and the phases of Venus. Finally, in 1632, he put all of his conclusions into a major work: the *Dialogue concerning the Two Chief World Systems, Ptolemaic and Copernican.*

By this time, Bellarmine was some twelve years dead. But the Inquisition was still alive and active, so the *Dialogue* is framed as a hypothetical discussion—an argument among three friends as to whether the heliocentric, geokinetic model could, theoretically, prove to be the best possible picture of the universe. The Copernican model is defended by the thoughtful and intelligent characters Salviati and Sagredo; all Inquisition-approved opinions are voiced by the least sympathetic character, the clearly ignorant and incompetent Simplicius, blindly loyal to Aristotle, willing to check his reason at the door.

The ruse was sufficient to get the *Dialogue* past the initial censor, the Dominican theologian Niccolo Riccardi, although Riccardi insisted on a preface that recognized the church's perfectly valid objections to heliocentrism. He also wanted a disclaimer at the end cautioning that the tides could be understood without recourse to a moving earth. Galileo promptly supplied a highly sarcastic preface ("Several years ago there was published in Rome a salutary edict which, in order to obviate the dangerous tendencies of our present age, imposed a seasonable silence upon the . . . opinion that the earth moves"), and placed in Simplicius's mouth an ending assertion that God, "in His infinite power and wisdom," was probably causing the tides to move "in many ways which are unthinkable to our minds."[13]

The preface seems to have satisfied Riccardi, who was not a sophisticated reader, and in February 1632 a thousand copies of the *Dialogue* were printed in Florence. The print run immediately sold out. By early spring, churchmen who had better ears for irony than Riccardi were beginning to point out that Galileo had undoubtedly violated Bellarmine's earlier warning.

Galileo retorted that the *Dialogue* was clearly hypothetical, and that he was neither holding nor defending heliocentrism—merely discussing it (which Bellarmine had explicitly allowed him to do). But the current head of the Inquisition, Cardinal Vincenzo Maculano, disagreed. In the spring of 1633, Galileo was forced to travel to Rome to defend himself. Maculano continued to be unimpressed by his arguments, and on April 28 he threatened the old man with "greater rigor of procedure" if he would not confess that he had broken Bellarmine's strictures.

Greater rigor of procedure: This was a code phrase for torture, and Galileo—now in his seventies and unwell—buckled. On June 22 he knelt in front of an assembly of churchmen and recited, obediently, "I abandon the false opinion which maintains that the Sun is the center and immovable." With this confession in hand, Maculano sentenced Galileo to house arrest, ordered him to recite seven penitential psalms once a week for three years, and banned the *Dialogue* forever.[14]

Under house arrest, Galileo went back to his studies on motion; in 1638 his work *Two New Sciences*, an exploration of non-Aristotelian physics, was printed in Leiden, where it did not have to receive the approval of a church censor. He died in 1642, his condemnation still active and the *Dialogue* still on the *Index*.

But outside the reach of the Inquisition, the *Dialogue* continued to circulate: reprinted, read throughout Europe, translated into English in 1661. Barely a quarter century after Galileo's death, the geocentric model was dead, and Aristotelian physics had been almost entirely superseded by new ways of thinking.

To read relevant excerpts from Dialogue concerning the Two Chief World Systems, *visit http://susanwisebauer.com/story-of-science.*

GALILEO GALILEI
Dialogue concerning the Two Chief World Systems
(1632)

The best and most readable English translation is by Stillman Drake. Originally published in 1953, it is available in a nicely revised and annotated edition from the Modern Library Science Series.

Galileo Galilei, *Dialogue concerning the Two Chief World Systems, Ptolemaic and Copernican*, trans. and with revised notes by Stillman Drake, Modern Library (paperback, 2001, ISBN 978-0375757662).

Instruments and Helps

*Improving the experimental method
by distorting nature and extending the senses*

I am not a little pleased to find that you are resolved, on this
occasion, to insist rather on Experiments than Syllogisms.
—Robert Boyle, *The Sceptical Chymist*, 1661

The next care to be taken, in respect of the Senses, is a
supplying of their infirmities with Instruments, and, as it
were, the adding of artificial Organs to the natural.
—Robert Hooke, *Micrographia*, 1665

In 1641, the year before Galileo's death, the Irish teenager Robert
Boyle set off on his "grand tour"—a trip through Europe in the company of his tutor, a rite of passage for well-to-do young men who
had just finished secondary school (in Boyle's case, Eton). By fall, the
two travelers had reached the north of Italy and decided to winter in
Florence. There, as Boyle's early biographer Thomas Birch tells us, he

> spent much of his time in learning of his governor (who spake it
> perfectly) the Italian tongue, in which he quickly attained a native
> accent, and knowledge enough to understand both books and
> men. . . . The rest of his spare hours he spent in reading the mod-
> ern history in Italian, and the new paradoxes of the great star-gazer
> Galileo, whose ingenious books . . . were confuted by a decree
> from Rome.[1]

Confuted or not, Galileo's works were clearly in wide circulation, and Robert Boyle found them entirely convincing: "This hypothesis of the earth's motion," he later wrote, "is far more agreeable to the phenomena, than the doctrine of Aristotle (who was plainly mistaken)."[2]

By this point, Galileo's telescopic observations from thirty years earlier had been duplicated, confirmed, and elaborated upon. Telescopes themselves had become stronger and better; the astronomer Johannes Kepler had theorized that two convex lenses would give a clearer, wider field of vision than the combination of convex and concave lenses used by Galileo, and in 1614 the German physicist Christoph Scheiner had built a Keplerian scope and proved the point (although the improved image was upside down). The same technology had been exercised in the opposite direction as well; Galileo himself had made a compound microscope, which he called an *occhialino* ("little spectacles"), and had used it to contemplate "an infinite number of small animals with enormous admiration." ("The flea is very horrible," he wrote to the Roman naturalist Federico Cesi in 1624.)[3]

Seven years later, back from his tour and pursuing his own scientific research in London, Robert Boyle enthused over these new technologies. "With bold telescopes I survey the old and newly discovered stars and planets," he wrote, "with excellent microscopes I discern, in otherwise invisible objects, the unimitable subtlety of nature's curious workmanship . . . [and] by . . . the light of chymical furnaces, I study the book of nature."[4]

Chymical furnaces: This was a general term for vessels (porcelain flasks, pottery kilns, water or sand baths, brick or stone ovens) in which natural substances could be heated to artificial temperatures in order to find out more about their properties. Telescopes, microscopes, and furnaces were essentially unlike the instruments that Pythagoras and Archimedes and Aristotle had used in their investigations of the natural world. The ancients had measured nature, weighed it, calculated it, using their senses to comprehend the physical world. But scopes and furnaces changed the basic relationship between the senses and the subject. They *distorted* the natural world—magnifying it unnaturally, fusing or melting or distilling it (in Boyle's own words, "torturing" natural bodies "into a confession of their constituent principles").[5]

This was part of the Baconian experimental program. "Neither the naked hand nor the understanding left to itself can effect much," Bacon had written in the *Novum organum*. "It is by instruments and helps that the work is done." Experiments done with these instruments and helps were "elaborate," carried out in "elaboratories"—a word appearing in seventeenth-century English for the very first time.

An elaboratory took time and money to construct, but Robert Boyle was independently wealthy and unencumbered by family, ideally situated to make use of the new instruments. He had, according to his acquaintance John Aubrey, a "noble laboratory and several servants (apprentices to him) to look to it. . . . He will not spare for cost to get any rare secret."[6]

Although he was better funded than most, Boyle was not the only young man experimenting with the new instruments and helps. In London he made the acquaintance of like-minded natural philosophers: "The cornerstones of the *invisible*, or (as they term themselves) the *philosophical college*, do now and then honour me with their company," he wrote to a friend in 1646, "men of so capacious and searching spirits, that school-philosophy is but the lowest region of their knowledge."[7]

This "invisible college" was not, as later conspiracy theorists suggest, an occult or secret society. It was a loose network of amateur philosophers who met in shifting and overlapping circles to share their discoveries and discuss Baconian methods. The particular circle that young Boyle seems to have been most involved with was dominated by a Moravian theologian named John Amos Comenius, who was (oddly) both pro-Baconian and firmly anti-Copernican, since he felt that observation was essential to science, but also that the Bible was perfectly clear about the universe being geocentric.[8]

By the early 1650s, Boyle's skills in experimental science seem to have outstripped those of his London network, and he moved first to his family lands in Ireland and then, searching for other like-minded philosophers, to Oxford. There he hired a poverty-stricken student named Robert Hooke to help him in his elaboratory.

Together, the two men—Boyle now approaching thirty, Hooke barely into his twenties—worked to construct various devices that Boyle could use to "torture" nature into revealing its secrets. By 1658 they had succeeded in building an air pump, a notoriously complicated instrument that had first been publicly demonstrated by the German physicist Otto von Guericke in 1654. Guericke had used the air pump to suck all the air out of a hollow bronze ball formed from two separate hemispheres; he had then hooked the hemispheres to two eight-horse teams, which were unable to pull them apart. Guericke's goal was to disprove a particular Aristotelian theory—that there was no such thing as "empty space" in the universe. Aristotle had argued that every *place* in the universe was occupied by *something* (a position later summed up as "Nature abhors a vacuum"). Guericke's demonstration was meant to remove *everything*, even tiny invisible particles of air, from inside the ball in order to create a place where *nothing* existed—another mortal wound to Aristotelian physics.[9]

Boyle's own experiments with the air pump had a slightly different purpose. It was clear that nature did *not* abhor a vacuum; now he wanted to know what happened to various phenomena in the absence of air. Together, he and Hooke put all sorts of things into a chamber and pumped the air out: marbles, weights, feathers, a ringing watch alarm (they couldn't hear it), gunpowder (hard to ignite), candles (they went out), a duck (it fainted), and several snakes (eventually, they died). The experiments led Boyle to his first publication, 1660's *New Experiments Physico-Mechanical: Touching the Weight of the Air and Its Effects*. In it, he theorized that air particles can be understood as

a heap of little bodies, lying one upon another, as may be resembled to a fleece of wool. For this . . . consists of many slender and flexible hairs; each of which may indeed, like a little spring, be easily bent or rolled up; but will also, like a spring, be still endeavouring to stretch itself out again.[10]

Just as wool can be compressed within a closed fist into a more compact bundle, so also can air be pushed into a smaller area by external pressure. And, just as the wool expands again when the fist is unclenched, so also will air "spontaneously expand or display itself towards the recovery of its former, more loose and free condition" when the pressure is eased. This quality Boyle called the "spring of the air," and he expressed it in a formula now known as Boyle's Law: when gas is compressed into a smaller volume, its pressure increases.[11]

The formulation of Boyle's Law is generally seen as a milestone in the development of modern physics, but Robert Boyle was not primarily interested in physics. He was interested in the composition of air, because he was interested in the elements that made up the natural world: "His greatest delight," John Aubrey observed, "is chymistry."

.

Chymistry: in the seventeenth century, a realm occupied only by magicians and metalworkers.

There was, in Boyle's world, no field of endeavor called "chemistry." Since ancient times, craftsmen had worked with precious metals and with dyes, which required some practical knowledge of chemical reactions. The Egyptians and the Greeks knew of fusion and filtration, crystallization and distillation. They knew how to use forges and furnaces to change the composition of their rough materials, and they were adept at using sulfur, arsenic, and mercury to change the colors of metals; arsenic, for example, turned copper white, "transforming" it (in the eyes of observers) into something completely different.[12]

These techniques were known as *chemia*, a word borrowed from Egyptian by the Greeks. In the following centuries, Arab craftsmen further developed *chemia* (in Arabic, *al-chemia*), speculating, as they did so, on the nature of the transformations taking place. In the ninth century, at least two Arab thinkers (the Baghdad mystic Jabir ibn Hayyan and the Persian physician Abu Bakr Muhammad ibn Zakariyya al-Razi) suggested that, rather than being composed of the four traditional elements, metals were made up of the additional elements mercury and sulfur (an idea hinted at in Aristotle's treatise *Meteorology*). In the thirteenth century, another Aristotelian, the Italian metallurgist Geber, suggested that matter was actually made up of something called *corpuscles*, rather than atoms. Atoms, by definition, could not be divided; but corpuscles could be penetrated by mercury, which mutated their internal structure into something else. If matter was made up of corpuscles, the transformation of some metals into others—copper into silver, or lead into gold—could actually take place.[13]

This was a serious scientific theory, but the possibility that a worthless metal could be turned into gold opened up the practice of *al-chemia* to a host of tricksters and con men, who became adept at faking transformations and passing false "gold" off to gullible buyers. A partial rehabilitation of alchemy's reputation came in the hands of the sixteenth-century German doctor Paracelsus, who was interested in alchemical techniques as a way of producing better medicines, and who argued that all natural changes (growth and development, fermentation and digestion) were essentially alchemical. But Paracelsus, who suggested that Aristotle's four elements be replaced by three principles (the sulfur and mercury of the ninth-century alchemists, plus the additional principle of salt), was on the one hand a difficult, egomaniacal personality and, on the other, narrowly focused on alchemy's uses for medical prescriptions; his theories made few inroads into the larger world of natural philosophy.

Boyle believed that alchemy—the study of matter's makeup—had as much to contribute to natural philosophy as did physics or astronomy. And the work he did in his elaboratory suggested to him that neither the Aristotelian theory of four elements nor the Paracelsian system of three principles held up under repeated experimentation. According to Aristotle, fire would always convert other elements into fire; why, Boyle asked, does nature so often "miss her end" in his chemical furnaces?

> The flame does never turn the bricks that it makes red-hot into fire; nor the crucibles, nor the cupels* nor yet the gold and silver that it thoroughly pervades. . . . And even when fire acts upon wood, there is but one part of it turned into fire, since, to say nothing of the soot and concreted smoke, the ashes remain fixed and incombustible.[14]

---

*Cups made of bone-ash, used to combine molten metals.

As for Paracelsus's three principles, Boyle doubted that sulfur could be considered a "primeval element," since he had been able to produce a "sulphureous liquor" in his laboratory using, as ingredients, distilled liquids that were generally accepted by "chymists" as containing no sulfur at all.[15]

In 1661, Boyle published his second major work, *The Sceptical Chymist*. It was constructed, very traditionally, as a dialogue among four characters: Themistius, a disciple of Aristotle; Philoponus, a Paracelsian; Carneades, Boyle's mouthpiece for his own point of view; and Eleutherius, an interested bystander. Eleutherius asks well-informed questions; Themistius and Philoponus argue for the four-element and three-principle schemes, respectively; and Carneades demolishes them.

But *The Sceptical Chymist* is entirely nontraditional in the proofs that Carneades offers. He does not argue (as Aristotle or Copernicus would have) on the basis of having come up with a better, more coherent answer. In fact, although Carneades suggests a different theory of matter (a version of Geber's "corpuscles," a "universal matter" that clumps together into masses that can be split, altered, and transformed), he is aware that he has no proof.

But what he *does* have is a weight of repeated, elaborated experiments that *disprove* the Aristotelian and Paracelsian positions. Eleutherius, the intelligent layman, is delighted by this. "I am not a little pleased," he says to Carneades, approvingly,

> that you are resolved on this occasion to insist rather on experiments than syllogisms. For I, and no doubt you, have long observed, that those dialectical subtleties that the schoolmen too often employ about physiological mysteries, are wont much more to declare the wit of him that uses them, than increase the knowledge or remove the doubts of sober lovers of truth.[16]

It is not in syllogism—in coherent, unified *systems*—that truth will be found, but in repeated experimentation.

Boyle refers to experimental proof nearly 150 times throughout the pages of *The Sceptical Chymist*. Furthermore, as he points out in the preface, these are experiments that have been *performed* in an elaboratory—not merely set up as thought problems. Boyle was continually exasperated by the habit, which many of his contemporaries had developed, of philosophizing about the material world on the basis of "chymical experiments, which questionless they never tried; for if they had, they would . . . have found them not to be true." He was particularly irked by his contemporary Blaise Pascal's *Physical Treatises*, which claimed to base its conclusions about the behaviors of liquids on experimentation, when in fact Pascal's "trials" were merely "mental" experiments: "They require brass cylinders, or plugs," Boyle complained, "made with an exactness that, though easily supposed by a mathematician, will scarce be found obtainable by a tradesman." This was not good enough. Truths could not be discovered merely through the exercise of reason: "intricate and laborious experiments" had to be *done*. Which is why Boyle's preface also warns the reader "not to be forward to believe chymical experiments . . . unless he that delivers them mentions his doing it upon his own particular knowledge."[17]

And even then, the experiments should be done again, and then repeated yet *again*. "Try those experiments very carefully and more than once, upon which you mean to build considerable superstructures either theoretical or practical," Boyle later wrote to his readers, "and . . . think it unsafe to rely too much upon single experiments." Variations in conditions, or materials, could drastically affect the outcome of an experiment. Only results that could be replicated over and over again should serve as the basis for a theory.[18]

Bacon had laid the foundation for the modern scientific method, but Boyle's use of instruments and helps truly brought the experimental phase of modern science into being. *The Sceptical Chymist*'s place in history is assured not by its conclusions, but by its procedures; not by the truth it discovers at the end, but by the road it takes to get there.

·

In 1662, the year after *The Sceptical Chymist*'s publication, Boyle's lab assistant Robert Hooke acquired a new job: curator of experiments for the fledgling Royal Society of London.*

*The early history of the Royal Society of London is somewhat obscure and is also much debated. The accuracy of the earliest account, Thomas Sprat's *The His-*

Since the 1640s, as Boyle's early reference to the "invisible college" reveals, small groups of natural philosophers had been meeting informally in both London and Oxford: "They had no Rules nor Method fix'd," says Thomas Sprat, the first historian of the Royal Society, "[and] their intention was more . . . to communicate to each other . . . their discoveries . . . than a united, constant, or regular inquisition." This lack of structure fitted the chaotic, uncertain years of the English Commonwealth: a time when the English government was turned onto its head, when Boyle was best off researching quietly, and privately, away in his self-funded lab.*

But Oliver Cromwell, prime mover and lord protector of the commonwealth, died in 1658; Charles II returned triumphantly to his throne, and the traditional English hierarchies were restored. Six months after Charles II's recoronation, the Royalist scholar Christopher Wren—architect, astronomer, and physicist—chaired the first meeting of the newly formed "College for the Promoting of Physico-Mathematical Experimental Learning" in London. By 1662, the "college" had been granted a charter by Charles II and renamed the Royal Society of London.[19]

Now a royally favored institution, the Royal Society suddenly grew; in fact, its nonscientific fellows soon outnumbered the natural-philosopher types. Hooke's appointment as curator of experiments seems to have been intended to keep the group on track. He was paid a full-time stipend to do two things: present a variety of weekly experiments to the gathered Royal Society, explaining and demonstrating as he went; and assist the fellows with their own experiments, as needed.[20]

---

tory of the Royal Society of London (1667), has been challenged in recent years; Michael Hunter's Establishing the New Science: The Experience of the Early Royal Society (Boydell Press, 1989) offers a lively and useful survey of the evidence.

*The English Commonwealth (1649–60) was a brief and violent departure from the tradition of English monarchy. Parliamentary leaders, including the Puritan Oliver Cromwell, led a revolt against King Charles I that ended in the arrest, trial, and shocking execution of the sovereign. A nominally republican government took Charles I's place; within four years, Cromwell managed to seize personal control of England under the title "Lord Protector." The Commonwealth ended ignominiously with Cromwell's death, after which Charles I's exiled son and heir was invited back to the throne.

This paid position made Robert Hooke (probably) the first full-time salaried scientist in history. The Royal Society was made up of astronomers, geographers, physicians, philosophers, mathematicians, opticians, and even a few chemists, so Hooke was called on to experiment and research across the entire field of natural philosophy. The broad scope of the position suited him perfectly. His laboratory work with Boyle had touched only the surface of his abilities. He was an excellent mathematician (which Boyle was not), expert at grinding and using lenses, inventor of a barometer, founder (in some sense) of the modern study of meteorology, a competent geologist and biologist, architect and physicist.

The experiments that Hooke presented to the Royal Society, in its weekly meetings, stretched across the range of his abilities. One set of demonstrations, recorded by Thomas Birch, began with an experiment showing how fluids reach hydrostatic balance, continued with a Parisian method for waterproofing calf hide by soaking it in "salad oil," and concluded with the presentation of a "sort of Portugal onion" that all examined with interest, since "none of the like" had been seen before.[21]

Further experiments involved pendulums, distilled urine, insects placed in pressurized containers, observations through colored and plain glass, and the weight of water. But increasingly, Hooke's experimental demonstrations made use of the microscope.

The first work of natural philosophy to utilize the microscope had been published, some forty years earlier, by Federico Cesi, Galileo's correspondent in Rome. Called *Apiarium*, it was a thorough study of the habits of bees; Cesi had used microscopic observations both to confirm Aristotle ("Just as Aristotle wrote, so [we] saw and observed that they . . . carry away their pollen with these hairs") and to occasionally contradict him. Other microscopic studies had been done by Cesi's colleagues, and the lens technology had been slowly improving.[22]

Hooke was particularly interested in the microscope. Like his mentor Boyle, he believed that natural philosophy needed instruments: "By the addition of such artificial Instruments and methods," he wrote in 1664, "there may be, in some manner, a reparation made for the . . . infirmities of the Senses":

Some parts of [Nature] are too large to be comprehended, and some too little to be perceived. And from thence it must follow, that not having a full sensation of the Object, we must be very lame and imperfect in our conceptions about it . . . hence, we often take the shadow of things for the substance, small appearances for good similitudes, similitudes for definitions. . . . [The] care to be taken, in respect of the Senses, is a supplying of their infirmities with Instruments, and, as it were, the adding of artificial Organs to the natural. . . . By the help of Microscopes, there is nothing so small, as to escape our inquiry; hence there is a new visible World discovered to the understanding.[23]

Hooke had been demonstrating this "new visible World" to the fellows, and they were appreciative of his advances. In April of 1663, the minutes of the Royal Society note, "Mr. Hooke was charged to bring in at every meeting one microscopical observation at least." At the very next meeting, Hooke "showed the company" what moss looks like under magnification. "He was desired to continue," the minutes explain; and over the course of the next few months, Hooke demonstrated the microscopic structures of cork, bark, mold, leeches, spiders, and "a curious piece of petrified wood" belonging to one of the fellows, a Dr. Goddard.[24]

Dr. Goddard had asked him to examine and report on this odd specimen, and Hooke duly carried out his duties. The petrified wood, he wrote, resembled living wood in its pores and structure, but was as hard and impenetrable as rock. And then Hooke offered an explanation:

The reason . . . seems to be this: that this petrified wood having lain in some place, where it was well soaked with petrifying water (that is, such a water as is well impregnated with stony and earthy particles) did by degrees separate, by straining and filtration, or perhaps by precipitation, cohesion, or coagulation, abundance of stony particles from that permeating water; which stony particles . . . conveyed themselves . . . into the pores. . . . By this intrusion of the petrified particles, it also becomes hard and friable . . . the smaller pores of the wood being perfectly stuffed up with the stony particles.[25]

He had gone beyond observation with instruments, beyond the disproving of current theories, to something new: the establishment of a new physical process that he had not seen (and could not see), but that he was able to *deduce*. He had described, for the first time, the process of fossilization, and in doing so had contradicted the received wisdom—that fossils were nonorganic forms, produced and made out of rock.

This was the next step in the use of instruments and helps: using these closer observations, the extension of the senses through artificial means, as the launching place for new ways of *thinking*. Not only the senses, but also the *reason*, were augmented by these instruments and helps. The true end goal of the microscope and telescope, the air pump and the vacuum chamber, was not merely observation; it was observation that led to new theories, observation that allowed man's mind to range farther afield than ever before.

In 1664, the Royal Society formally requested that Hooke print his micrographic observations. On top of his other competencies, Hooke was a skilled draftsman and artist. Rather than merely describing his discoveries in words, or commissioning nonscientists to produce his drawings, he made his own: large, exquisitely detailed, and perfectly clear.

The resulting work, *Micrographia*, was published in 1665. The first fifty-seven illustrations and observations are microscopic; the last three, of refracted light, stars, and the moon, are telescopic. The quality of the illustrations was light-years ahead of anything that had been seen before, and the book was an instant sensation.

But although the eye-grabbing pictures attracted the most attention, even more notable is that, throughout, Hooke uses his newly extended senses to build new theories. After carefully examining the colors and layers of muscovite ("Moscovy-glass"), he goes beyond his observations to suggest nothing less than a theory of how light works: it is, he speculates, a "very short vibrating motion" propagated "through an Homogeneous medium by direct or straight lines."[26]

Like his theory of fossilization, this model could not be directly proved. But Hooke was merely carrying out the method he had laid out in *Micrographia*'s "Preface." It was not enough merely to extend the senses by way of instruments; reason must follow the path laid out by these observations, interpret them, and then check itself again. Using William Harvey's circulatory system as his analogy, Hooke explained that true natural philosophy

> is to begin with the Hands and Eyes, and to proceed on through the Memory, to be continued by the Reason; nor is it to stop there, but to come about to the Hands and Eyes again, and so, by a continual passage round from one Faculty to another, it is to be maintained in life and strength, as much as the body of man is by the circulation of the blood through the several parts of the body, the Arms, the Feet, the Lungs, the Heart, and the Head. If once this method were followed with diligence and attention, there is nothing that [does not lie] within the power of human Wit. . . . Talking and contention of Arguments would soon be turned into labours; all the fine dreams of Opinions, and universal metaphysical natures, which the luxury of subtle brains has devised, would quickly vanish, and give place to solid Histories, Experiments and Works. And as at first, mankind fell by tasting of the forbidden Tree of Knowledge, so we, their posterity, may be in part restored by the same way, not only by beholding and contemplating, but by tasting too those fruits of natural knowledge, that were never yet forbidden.[27]

Instruments and helps were no longer merely extensions of the senses; they had become, for Hooke, the Tree of Knowledge, the path to perfection.

To read relevant excerpts from The Sceptical Chymist *and* Micrographia, *visit http://susanwisebauer.com/story-of-science.*

ROBERT BOYLE
The Sceptical Chymist
(1661)

The Sceptical Chymist is not an easy read ("prolix, repetitive, disjointed, and occasionally contradictory," in Lawrence Principe's words).[28] However, the Preface, Physiological Conditions, and First Part give an excellent overview of Boyle's preoccupations and his commitment to experimentation.

Robert Boyle, *The Sceptical Chymist: The Classic 1661 Text*, Dover Publications (paperback and e-book, 2003, ISBN 978-0486 428253).

ROBERT HOOKE
Micrographia
(1665)

Although multiple reprints of the *Micrographia* are available, few of them reproduce Hooke's groundbreaking illustrations at the original size or with decent detail. The best way to view the illustrations is in the Octavo CD, which offers clear scans of the actual pages of the original book, in PDFs that can be magnified, rotated, and viewed in color or black and white.

Robert Hooke, *Micrographia*, Octavo Digital Rare Books (CD-ROM, 1998, ISBN 1-891788-02-7).

However, the text itself (complete with unmodernized spelling) is extremely difficult to make out in the Octavo scans. Consider turning to one of the free e-book versions (such as that found at Project Gutenberg) or a paperback reprint in order to read Hooke's accompanying essays—particularly the Preface, in which he explains the relationship between the senses and the faculty of reason.

Robert Hooke, *Micrographia*, Project Gutenberg (e-book, 2005).
Robert Hooke, *Micrographia*, Cosimo Classics (paperback, 2007, ISBN 978-1602066632).

TWELVE

Rules of Reasoning

Extending the experimental method
across the entire universe

Those qualities of bodies . . . that belong to all bodies on
which experiments can be made should be taken as qualities
of all bodies universally.

—Isaac Newton, *Philosophiae naturalis*
principia mathematica, 1687/1713/1726

Five years after the publication of the *Micrographia*, at the December
21, 1671, meeting of the Royal Society, Robert Hooke (still curator of
experiments) presented a new method of musical notation and proposed an experiment measuring the force it would take to make mercury pass through wood. Robert Boyle, also in attendance, described
an experiment showing that "air will flow where water will not."
And a new candidate for membership was proposed: one "Mr. Isaac
Newton, professor of mathematics in the University of Cambridge."[1]

Isaac Newton was twenty-nine years old, Cambridge educated,
and for the prior four years had been a teaching fellow at Trin-
ity College, his alma mater. He was duly elected at the January
1672 meeting, where "special mention was made of Mr. Newton's
improvement of telescopes"; Newton had sent the Royal Society
one of his improved instruments, with his compliments.

Newton was, like Boyle and Hooke, a user of artificial helps,
a mathematician who was also anxious to extend his senses in the
pursuit of truth. In the February meeting, the ink on his membership certificate barely dry, he submitted to the society a letter
describing his most recent "philosophical discovery":

that light is not a similar, but a heterogeneous body, consisting of different rays, which had essentially different refractions, abstracted from bodies through which they pass; and that colours are produced from such and such rays, whereof some, in their own nature, are disposed to produce red, others green, others blue, others purple . . . and that whiteness is nothing but a mixture of all sorts of colours, or that it is produced by all sorts of colours blended together.[2]

Newton had been working not with scopes, but with prisms: instruments that distorted ("tortured") natural light into revealing its component parts.

In his letter, Newton characterized his discovery on light as "the oddest, if not the most considerable detection, which hath hitherto been made in the operations of nature." This immodest claim immediately produced an equal but opposite reaction. The Royal Society "solemnly thanked" Mr. Newton for his discoveries and commissioned Hooke to produce an answer; Hooke fired back that he could "see no necessity" in Newton's proposal. It went directly against the commonly received wisdom that light was white, and homogenous. Passing it through various substances (such as transparent layers of stone, as in an experiment that Hooke had documented in *Micrographia*) merely modified it so that it turned different colors. Hooke conceded that Newton's new explanation *might* be true, but he insisted that it had no *greater* probability of truth than the existing theory; Newton, he objected, had not provided "an absolute demonstration of his theory," and Hooke declined to be convinced. Later that year, he duplicated Newton's experiment for the Royal Society, "confirming what Mr. Newton had said," but he continued to argue that "these experiments were not cogent to prove that light consists of different substances." He could think of at least two other "various hypotheses" that could equally well explain Newton's results.[3]

It was the beginning of a contentious relationship between the two men. The enmity grew partly out of a natural clash between two highly competent and ego-driven personalities, and partly out of a difference in philosophy. Hooke, along with most of the leading members of the Royal Society, was entirely committed to the experimental method, but wary about drawing universal conclusions from it. The society was intrigued by Newton's "theory of light" but recommended that many more experiments be carried out before any conclusions could be drawn. These experiments dragged on for the next three years, with much correspondence flying back and forth between Newton's Cambridge elaboratory and the Royal Society's London headquarters.

By 1675, Newton was growing exasperated. He sent a much more elaborate manuscript to the society, laying out a more detailed explanation about his experiments and what they revealed about light movement through the "ether." On December 16, Hooke presented this manuscript to the Royal Society: "After reading this discourse," the minutes tell us, "Mr. Hooke said that the main of it was contained in his *Micrographia*, which Mr. Newton had only carried farther in some particulars."

Newton, unsurprisingly, took offense and, in return, accused Hooke of "borrowing" a little too freely from other thinkers, including both René Descartes (who had published his *Discourse on Method* forty years earlier) and Newton himself. The spat between the two grew sharp and public. More letters were exchanged, and more experiments proposed. Newton became increasingly frustrated—partly by Hooke, but mostly by the Royal Society's continuing demands for more proof. In 1676 he observed tartly in a response to the society itself, "It is not number of experiments, but weight, to be regarded; and where one will do, what need many?" And he wrote bitterly, to a colleague, "I see that a man must either resolve to put out nothing new, or to become a slave to defend it."[4]

Gradually, Newton withdrew from participation in the Royal Society's agenda. His name appears less and less often in the minutes. Rather than participating in the unending experiments curated by Hooke and demanded by the society's fellows, Newton devoted himself to his own research: not only on light and optics, but also on the orbits of the planets and the celestial mechanics that might explain them. Twelve years after his fracas with Hooke, he published his first major work: *Philosophiae naturalis principia mathematica*, or "Mathematical Principles of Natural Philosophy."

The *Principia* was concerned, mostly, with Galileo's heliocentric model: the orbit of the planets around the sun. Increasingly accepted as consistent with reality, Galileo's solar system was still plagued with questions. Galileo himself had solved some of its major problems, but by no means all of them. He believed that heliocentrism was the best explanation for the movement of the tides, and his telescopic observations had revealed that it was possible for planets to rotate around heavenly bodies that were *not* the earth. His experiments with weights, whether or not dropped from the Leaning Tower, had shown that the earth could be rotating without objects flying off its surface.

But he had not tried to explain *how* the force that caused weights to drop (the *gravitas*, or "heaviness") worked. In fact, although he had disproved some of the central aspects of Aristotelian physics, he had held on to the vaguely Aristotelian idea that *gravitas* was an intrinsic property of physical objects, not an outside force that affected them.[5]

And Galileo could not explain why, if planets orbited the sun, calculations about their circular orbits didn't account for their movements. His contemporary Johannes Kepler had proposed laws for *elliptical* orbits that yielded much better results—but neither Kepler nor Galileo had been able to explain *why* the orbits should be elliptical rather than circular. In the Galilean universe there was no force, and no property possessed by physical bodies, that would propel a planet into an ellipse rather than a circle.

This was the presenting problem that Newton solved.

He did so by extending the results of Galileo's earthly experiments with weights into the heavens. Galileo had theorized that the *gravitas* of objects meant they would continue to travel at the same rate, no matter how far they fell; Newton suggested that *gravitas* was not an inherent quality, but a force, exerted by the sun on the planets and by planets on the moons surrounding them. The same *gravitas* that drew Galileo's objects to the earth also drew the moon toward the earth—but the strength of this force did not remain the same over distance. It *changed*. As the planets moved farther from the sun, the force that pulled on them weakened—thus, the ellipse.[6]

To fully explain the laws governing this new force—most vitally, the relationship of the distance between two objects and the strength of the *gravitas* between them—Newton had to come up with an improved mathematics, capable of accounting for continual small changes. This new math was a "mathematics of change," able to predict results in a setting where conditions were constantly shifting, forces altering, factors appearing and receding.[7]

So, the *Principia* performed two groundbreaking tasks simultaneously. It explained the *why* behind the ellipses of the planets—and, in doing so, revealed for the first time a new force in the universe: the force of gravity. It also introduced an entirely new branch of mathematics; a dynamic mathematics that became known in the seventeenth century as *calculus*: from the Latin word for "pebble," one of the tiny stones used as arithmetical counters.*

•

The four books of the *Principia* lay out the rules by which gravity functions. Throughout, Newton establishes and makes use of three principles (Newton's Laws of Motion):

Law	Formula	Nonmathematical statement	Paraphrase
1. The Law of Inertia	$\sum F = 0 = \dfrac{dv}{dt} = 0$	"The velocity of an object remains constant unless an unbalanced force acts on the object."	Objects in motion remain in motion, and objects at rest remain at rest—unless an outside force is applied.
2. The Law of Acceleration	$F = m\dfrac{dv}{dt} = ma$	"The net force on an object is equal to its mass times its acceleration and points in the direction of the acceleration."	When a force is applied to a mass, acceleration results. The greater the mass, the greater the force needed to produce acceleration.

* Isaac Newton and his contemporary Gottfried Leibniz were simultaneously, and independently, working toward this new "calculus." Afterward, they fought bitterly about who had invented which aspects of calculus, and who had copied from whom; this quarrel takes up a lot of literature about Newton, but it is irrelevant to our interests here. A useful overview is found in Chapter 15 of Niccolò Guicciardini's *Isaac Newton on Mathematical Certainty and Method* (MIT Press, 2009).

3. The Law of Action and Reaction	$F_A = -F_B$	"If an object exerts a force on a second object, the second object exerts an equal force back on the first object."	For every action, there is an equal and opposite reaction.

Nonmathematical statements are all from Larry Kirkpatrick and Gregory Francis, *Physics: A World View* (Thomson, 2007), 37, 41.

Books I and II establish these laws of motion, both in the abstract (without any friction present) and in the presence of resistance; the third book discusses gravity as a universal force.

None of this is easy going. The title of the *Principia* points to the book's greatest difficulty: it is composed largely of impenetrable mathematical explanations, laid out in Newton's new calculus. William Derham, Newton's longtime friend and colleague, later wrote that the book's obscurity was deliberate. Newton had confided in him that, since he "abhorred all Contests," he "designedly made his Principia abstruse," in order to "avoid being baited by little Smatterers in Mathematicks." Since "little smatterers" is probably an accurate description for the majority of interested modern readers, including myself, the bulk of the *Principia* remains out of reach for most of us. (This isn't necessarily a failing of modern education. As James Axtell has observed, Newton's strategy "effectively rendered the *Principia* unintelligible . . . to the virtuosi and intellectual laymen" of his own day as well; a frustrated Cambridge undergraduate famously remarked, as Newton passed him on the street, "There goes the man who has writt a book that neither he nor any one else understands.")[8]

But there are two sections of the *Principia* in which Newton abandons his dense formulaic prose and writes clearly: the "General Scholium" (which is to say, "Overall Explanatory Remarks") at the very end of the entire *Principia*, and the "Rules for the Study of Natural Philosophy" that come at the beginning of Book III.

The "Rules" are, in a way, Newton's final response to the Royal Society. He was aware that the conclusions of the *Principia* could be dismissed by the literal-minded as "ingenious Romance"—mere guesses, airy speculations. After all, he had not actually experimented with the moon, or spun planets at different distances from the sun to observe the rates of their orbits. Instead, he had taken the results of experiments carried out on the earth and had extrapolated their results into the heavens—a method that the pedantic Royal Society might not applaud.[9]

The "Rules" explain why Newton's conclusions about the movements of the moon and planets, while not experimentally proven in the way that would make Hooke happy, are nevertheless reliable. The first three are:

1. Simpler causes are more likely to be true than complex ones.
2. Phenomena of the same kind (for example, falling stones in Europe and falling stones in America) are likely to have the same causes.
3. If a property can be demonstrated to belong to all bodies on which experiments can be made, it can be assumed to belong to all bodies in the universe.

The "Rules" do not appear in exactly this form in the first edition of the *Principia*, although Newton's conclusions make clear that he was certainly operating with them in mind. Not until the second edition, published in 1713, was he able to put his working assumptions into words. And he did not add his fourth and final rule until the third edition of the *Principia*, in 1726:

4. A general theory that is based on specific phenomena or experimental results should be considered true unless *new* phenomena or *additional* experimental results make another theory more likely.

This is Bacon's inductive reasoning, always progressing from specifics to generalities, but extended by Newton to breathtaking lengths: across the entire face of the universe.

But the "General Scholium" (which also contains a famous discussion of the place of God in natural philosophy) places limits on the method. Gravity, Newton explains, is a force that penetrates as far as the centers of the sun and planets without any diminution of its power to act, and that acts not in proportion to the quantity of the *surfaces* of the particles on which it acts . . . but in proportion to the quantity of *solid* matter, and whose action is extended everywhere to immense distances, always decreasing as the squares of the distances.[10]

But, he cautions, "I have not yet assigned a cause to gravity." He could deduce the laws of gravity from his experiments on the earth, but the *reason* for gravity lay beyond his grasp.

To go from laws to cause was, in Newton's view, theorizing in the absence of proofs—the sort of grand paradigm-inventing carried out by the ancient philosophers. He calls this, scornfully, "feigning the hypothesis": "I have not as yet been able to deduce from phenomena the reason for these properties of gravity," he concludes, "and I do not 'feign' hypotheses." He felt no need to provide a universal explanation for *why* the universe functions as it does—a theory of everything. In his "experimental philosophy,"

propositions are deduced from the phenomena and are made general by induction. The impenetrability, mobility, and impetus of bodies, and the laws of motion and the law of gravity have been found by this method. And it is enough that gravity really exists and acts according to the laws that we have set forth and is sufficient to explain all the motions of the heavenly bodies and of our sea.[11]

It is enough: with this, Newton was content. He had extended the reach of the experimental method across the universe, but he had also erected a boundary fence on its far side.

To read relevant excerpts from the Principia, *visit http://susanwisebauer .com/story-of-science.*

ISAAC NEWTON
"Rules for the Study of Natural Philosophy" and "General Scholium" from *Philosophiae naturalis principia mathematica* (1687/1713/1726)

Readers who want to tackle the entire *Principia* have several options. The clearest modern translation, done by I. Bernard Cohen and Anne Whitman, is available in a massive 950-page paperback; the entire first half of the book is commentary, explanation, and how-to-read-this-difficult-book guidance.

Isaac Newton, *The Principia: Mathematical Principles of Natural Philosophy: A New Translation*, trans. I. Bernard Cohen and Anne Whitman, assisted by Julia Budenz, University of California Press (paperback, 1999, ISBN 978-0520088177).

Multiple editions of the 1729 translation by Andrew Motte are also available. Although dated, and in places inaccurate, the prose is not significantly more difficult than that of the Cohen and Whitman translation.

Isaac Newton, *The Principia: Mathematical Principles of Natural Philosophy*, trans. Andrew Motte, Daniel Adee, publisher (free e-book, 1846).
Isaac Newton, *The Principia: Mathematical Principles of Natural Philosophy*, trans. Andrew Motte, Snowball Publishing (e-book and paperback, 2010, ISBN 978-1607962403).

Selected excerpts from the *Principia* (including the "Rules" and parts of the "General Scholium") and many other Newtonian writings, along with commentary, can be found in the Norton Critical Edition of Newton's work.

I. Bernard Cohen and Richard S. Westfall, eds., *Newton: Texts, Backgrounds, Commentaries*, W. W. Norton (paperback, 1995, ISBN 978-0393959024).

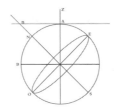

THIRTEEN

The Genesis of Geology

The creation of the science of the earth

> The general history of the earth ought to precede that of its productions.
>
> —Georges-Louis Leclerc, Comte de Buffon,
> *Natural History: General and Particular,* 1749–88

While physics and astronomy were flourishing, the study of the earth had remained mostly geographic.

The Greeks (of course) had created geography. Maps had been drawn since the days of the ancient Babylonians, but the conquests of Alexander the Great in the fourth century BC had suddenly opened up new possibilities to Greek mapmakers. A hundred years later, the librarian at Alexandria, Eratosthenes, wrote the first scholarly study of the earth's topography. The three volumes of his *Geographika* (the earliest known use of the word) laid out a new, worldwide system of parallels and meridians. Not long after, the Greek astronomer Hipparchus used his observations of the moon to create the grid of lines that we now know as latitude and longitude, improving the accuracy of Greek-created maps even further.[1]

So earth science began. But Greek geography simply observed the present state of the earth's surface. It offered no explanations about how the planet had come to be as it was, or why it functioned as it did. After all, Aristotelian philosophy held that the history of the earth was an infinite one, in which cycles of time were repeated ad infinitum; this did not suggest to the Greeks that research into the origin of the globe was even possible, let alone that it would be useful in understanding the world's present form.[2]

Advances in chemistry and in physics provided some useful insights about physical processes taking place *on* the earth. But it was not until the science of the stars had begun to develop toward greater maturity that the earth *itself* (extraordinary, but now seen to be one heavenly body among many; sharing qualities with the wandering planets, but unique as the home of the human race) became an object of study. Geology, as Charles Van Hise observes, was the child of astronomy.[3]

During the first part of the seventeenth century, though, geology was not quite ready to take its place as a new science. It was merely another branch of natural philosophy, still wondering what questions to ask. And those first questions were asked not by astronomers and physicists, but by theologians and philosophers.

In 1647 the natural philosopher John Lightfoot used the genealogical accounts of the Old Testament to calculate the age of the earth. It had been created, he announced, in September of 3928 BC. Three years later, the Irish bishop and amateur astronomer James Ussher merged the Bible chronologies with his own astronomical observations, and placed the time of creation slightly earlier. "In the beginning God created the heaven and the earth," Ussher wrote, in his *Annals of the World*. "The beginning of time, according to our chronology, happened at the start of the evening [midnight] preceding the 23rd of October, 4004 BC."[4]

This was not exactly science. Still, the Christian and Hebrew tradition that the earth's life span began at the moment of creation was an advance over the Greek tradition of endlessly repeating cycles. Assigning the earth a real, linear, chronological history suggested that its rocks and soil, its mountains and valleys, bore tracks of the actual past.

Nineteen years after Ussher set his date, the Danish clergyman Nicholas Steno, trained in anatomy at the universities of Copenhagen and Amsterdam before his ordination, published an essay about that past—the first real attempt at earth science.

The work, "Preliminary Discourse to a Dissertation on a Solid

Body Naturally Contained within a Solid," dealt with the puzzle of fossils: stone formations, often found in mountains far removed from the sea, that resembled living creatures. The first-century Roman philosopher Pliny had guessed that they were rocks fallen from the sky. The medieval philosopher and physician Avicenna suggested that they had been formed, mysteriously, by a "plastic" force in the earth that molded rock into new shapes. Robert Hooke, eyeing them under a microscope, had concluded that fossils were petrified remnants of living organisms, but he had been unable to explain why they were found in the midst of solid earth.[5]

Steno took Hooke's conclusions a little further. In 1666, Steno had carried out a careful dissection of the head of a strange shark, caught off the Italian coast and sent to him by an influential acquaintance, Ferdinando II de' Medici. The shark's teeth were identical to small tooth-shaped fossils that Steno and other seventeenth-century natural philosophers knew well: *glossopetrae*, stones resembling tiny forked serpent's tongues that had been gathered at numerous locations across Europe. The "Preliminary Discourse" aimed, first of all, to prove that *glossopetrae*, like the shark's teeth, were parts of previously living animals; and second, to explain how they came to be contained in stone.

Steno's explanation for the latter is obscured by clarifications, qualifications, unnecessary details, and unpruned language. But it can be pared down to three essential principles:

Steno's Principle of Superposition. Layers of rock are formed when rock particles settle at the bottom of water and are then compressed. So, lower layers of rock and artifacts were deposited earlier than higher layers.

Steno's Principle of Original Horizontality. Layers of rock always form horizontally. If a layer slants, or runs vertically, it was pushed that way by later factors.

Steno's Principle of Lateral Continuity. Layers of rock do not just end. So, if two layers of rock are near each other and have the same mineral content, soil content, and artifact content, but are separated from each other, they were originally the same layer but were disrupted by a later event.

All three of these principles point to the same deep insight, one that had never been articulated before: the layers of the earth, the *strata* (the "blankets" of rock and dirt that lie one on top of another) had been deposited, one at a time, over the course of many years. Digging down through them, the natural philosopher could travel back in time.

There was still no such thing as *geology*, but Steno had now made the invention of this science a possibility. He had identified the raw material that natural philosophers could exercise their Baconian methods upon: the earth's layers, and the fossils within them, were the *things* that could be observed, analyzed, theorized about.

At once, that raw material began to cast doubt on the conclusions of the theologians and philosophers.

Steno himself saw no difficulty with the creation date of 4004 BC. He was convinced that fossils were remnants of living creatures that had settled at the bottoms of streams, and it stretched credulity that fragile objects such as shells and claws (even petrified) could last for thousands of years. In the "Preliminary Discourse," Steno is clearly worried that a creation date of 4004 BC might actually be way too far back in the past.[6]

But other thinkers saw the raw material differently. Isaac Newton, paying more attention to the earth's core than to its fossils, had already speculated that the earth might originally have been a molten sphere. In that case the earth's age could be calculated by taking into account the rate at which iron cools (always keeping in mind that large spheres retain heat differently than smaller ones, since the relationship between the quickly cooling surface and the remainder of the sphere's mass changes). "And therefore," Newton concludes,

> a globe of red hot iron equal to our earth, that is, about 40,000,000 feet in diameter, would scarcely cool in an equal number of days, or in above 50,000 years. . . . I should be glad that the true proportion was investigated by experiment.[7]

In other words, if the earth *had* originally been molten, there was no way it was only 6,000 years old. But this was, for Newton, a "feigned" hypothesis, one that science did not yet have the tools to prove.

Newton's colleague and sometimes competitor, the German mathematician Gottfried Wilhelm Leibniz, offered a similar speculation—that the earth had once been liquefied, like metal, and had cooled and hardened over time. This process had produced large bubbles; some of them calcified into mountains, others shattered and disintegrated, producing valleys. But Leibniz was also wary of the feigned hypothesis, and he also declined to provide a possible age for the earth.[8]

The age of the earth, and how the layers of the globe might be interpreted in order to find it, might have remained an open and honestly debated question. But in 1701 the bishop of Winchester, William Lloyd, added James Ussher's date of 4004 BC to the marginal notes of the newest version of the 1611 Authorized Version of the Bible (the most widely read and influential English-language version of the Bible in print). This lent the appearance of sacred authority to Ussher's speculation; from this point on, proposing an age of more than 6,000 years for the earth would carry with it the slur of denying scripture.

It was fifty years before another scholar dared to suggest a longer chronology. Even then, the suggestion was made in the form of fiction, and underground fiction at that.

In the 1720s the French natural philosopher Benoît de Maillet circulated, privately, a set of dialogues between an Indian philosopher named Telliamed and a French missionary about the age of the earth.* Telliamed, pointing to centuries of measurements showing that the level of the Mediterranean was dropping, argued that the whole earth had once been covered in sea, but this sea was constantly seeping into a vortex at the globe's center. The rate of the water's fall meant that the earth was at least *two billion years old*.

But this estimate need not contradict scripture: "You look upon

---

*Church authorities probably noticed that the name Telliamed was merely "de Maillet" printed backward, but we have no record of their reaction.

[the question of the earth's age] to be necessarily connected with your religion," Telliamed tells the missionary, "though in my own opinion, the former is quite indifferent to the latter." The study of the earth, he explains, cannot be done through the lens of faith. Rather, the planet must be examined with the objectivity that the astronomer brings to any other phenomenon:

> Let us not measure the past duration of the world, by that of our own years. Let us carefully consider what presents itself to our view in this universe, this immensity of the firmament, where we see so many other stars like our own sparkling. . . . Let us imagine to ourselves what is rendered highly probable since the invention of telescopes, that if we were placed at the highest point of this distance from our earth, which we can reach with them, we should perhaps discover as many worlds above us, which would be no less distant from our view.[9]

Not until the earth was accepted as one heavenly body among a myriad of others could *geology* come into existence. And not until human years were dismissed as the measure of all things could the time frame that geology demanded be laid out.

•

In the middle of the eighteenth century, the Comte de Buffon began to build the new science.

Buffon had been born Georges-Louis Leclerc, son of a salt tax administrator. The tax man had inherited a massive fortune and used it to buy the nearby village of Buffon; this purchase made him brand-new nobility. Young Georges-Louis, sensitive over his humble origins, abandoned the surname Leclerc in his twenties and merely signed himself "Buffon" for the rest of his life.[10]

His family wealth allowed him to indulge his wide-ranging interests in mathematics, physics, chemistry, microscopics, and botany. He tinkered, investigated, wrote, lectured, published. By the time he was in his thirties, his impressively broad accomplishments had caught the eye of the French court. In 1739 he was given the post of curator of the royal gardens, in charge of expanding and

improving both the gardens and the zoo inside it; he held this position for the rest of his life.

Buffon's duties as curator tended to focus his wide-ranging
interests, more closely, onto the earth and its living systems. In 1740
he announced the beginnings of a massive encyclopedic work, a
fifty-volume *Natural History*. It was a hugely ambitious expansion of
the project that Aristotle had undertaken in the *History of Animals*,
encompassing not just animal life, but botany as well. And as Buffon began to work, he realized that he could not begin with either.
"The general history of the earth," he wrote, at the very beginning
of his project, "ought to precede that of its productions."[11]

So the first volume of the *Natural History, General and Particular*
deals with the globe: its internal structure, its "form and manner
of existence," and its history. Buffon proposed to treat these things
scientifically, inductively, *Baconically*. He rejected any explanation
that relied on a single extraordinary event in the past, one that
could be neither observed nor repeated, as "unstable" and "constructed on tottering foundations." (Here, Buffon had in mind a
popular theory, proposed by the mathematician William Whiston,
that the present form of the earth could be explained by a primeval
collision with the tail of a comet.)[12]

Instead, Buffon insisted that the science of the earth must adopt,
as causes, *only* physical processes that can still be observed: the
movement of water, the gradual cooling of heated substances, the
erosion of soil.

> I speak not here of causes removed beyond the sphere of our
> knowledge, of those convulsions of nature, the slightest throe of
> which would be fatal to the globe. The near approach of a comet,
> the absence of the moon, or the introduction of a new planet into
> the system, &c. are suppositions upon which the imagination may
> rove at large. Causes of this kind will produce any effect we
> choose. From a single hypothesis of this nature, a thousand phys-
> ical romances might be drawn, and their authors might dignify
> them with the title of Theory of the Earth. . . . I reject these vain
> speculations: They depend upon mere possibilities, which, if
> called into action, necessarily imply such a devastation in the uni-

verse, that our globe, like a fugitive particle of matter, escapes our observation, and is no longer worthy of our attention. But, to give consistency to our ideas, we must take the earth as it is, examine its different parts with minuteness, and, by induction, judge of the future, from what at present exists. We ought not to be affected by causes which seldom act, and whose action is always sudden and violent. These have no place in the ordinary course of nature. But operations uniformly repeated, motions which succeed one another without interruption, are the causes which alone ought to be the foundation of our reasoning.[13]

So, what "uniformly repeated" operations might have shaped the earth? With a nod back to Newton, Buffon proposed that the planet had begun its life as a molten globe, slowly cooling toward its present temperature.

Granted, he could not observe this particular bit of the earth's past with his own eyes. But temperature readings of the earth's surface and deep mines clearly showed that the core was hotter than the surface, and *this* could be explained through a physical process that *could* be repeated again and again: Buffon heated iron spheres of different sizes until they glowed white, and then measured the time it took them to cool. Then he extrapolated those results to a body the size of the earth and concluded that the initial cooling of the globe had begun to take place 74,832 years before. Privately, he thought that an even longer time frame was probable—perhaps as long as three billion years (which is not so far off from the contemporary estimate of 4.57 billion).[14]

Laying out these theories in Volume 1 of the *Natural History*, Buffon simultaneously rejected the possibility of extraordinary divine intervention in the earth's past and contradicted Ussher's chronology. The book was an eighteenth-century best seller. Copies sat on the tables of well-educated men and women all over France, England, Holland, and Germany. Prominent members of the French Academy of Sciences repeatedly attacked Buffon's conclusions. The Faculty of Theology in Paris carried on a long and suspicious correspondence with him over his understanding of Genesis.[15]

Buffon read the correspondence, but his commitment to his

hypothesis merely deepened. Thirty years later, when he published a series of corrections and supplements to the existing volumes of the *Natural History*, he included a more detailed explanation of the stages of the earth's formation, divided into seven periods:

First Epoch	The earth begins to cool.
Second Epoch	The earth solidifies.
Third Epoch	Water covers the earth.
Fourth Epoch	The waters begin to recede and volcanic activity begins.
Fifth Epoch	Elephants and "southern animals" inhabit the warm north.
Sixth Epoch	Continents separate.
Seventh Epoch	Human life begins.

These "Epochs of Nature" were the first "deep time" chronology of the earth—a phrase coined, centuries later, by John McPhee to describe the entirely different time frame (a million years are as a day) used in the study of geology.

The Epochs of Nature chronology was a bit startling to Buffon's English translator, William Smellie—a little too un-British for his taste—so he abridged that particular section of Volume 9. "As this theory, however it may be relished on the Continent, is perhaps too fanciful to receive the general approbation of the cool and deliberate Briton," he explained in a translator's note, "the translator has been advised not to render it into English."[16]

But Buffon made no apologies. His insistence that no *extraordinary* events be used as explanation—the first principle of geology—had led him, inevitably, to the second: the history of the earth was a long, long one.[17]

To read relevant excerpts from the Natural History, General and Particular, *visit http://susanwisebauer.com/story-of-science.*

GEORGES-LOUIS LECLERC, COMTE DE BUFFON
Natural History, General and Particular
(1749–88)

Volume 1 consists of only two major sections: Chapter One, "The History and Theory of the Earth," which lays out all of Buffon's theories about the earth's formation, and the very long Chapter Two, "Proofs of the Theory of the Earth," divided into nineteen "articles" that provide detailed arguments and experimental support. The first chapter is worth a close and careful read; you may wish to skim the second.

Volume 9 contains numerous additions and supplements to different sections of Volume 1. You need only read the section titled "Facts and Arguments in Support of the Count de Buffon's Epochs of Nature: Of Giants, of the Glaciers, of the North-East Passage, concerning That Period When the Powers of Man Aided Those of Nature."

William Smellie's translation remains the only English version; it is readable and entertaining, but remember that his section on the Epochs of Nature is a paraphrase. It can be read in numerous free e-book versions, but the easiest to find is hosted by the University of Michigan's Eighteenth Century Collections Online (see http://susanwisebauer.com/story-of-science for a live link).

Georges-Louis Leclerc, Comte de Buffon, *Natural History: General and Particular*, trans. William Smellie, vol. 1 (e-book, 1780, no ISBN).

Georges-Louis Leclerc, Comte de Buffon, *Natural History: General and Particular*, trans. William Smellie, vol. 9 (e-book, 1785, no ISBN).

FOURTEEN

The Laws of the New Science

Two different theories are proposed
as explanations for the earth's present form

The production of our present continents must have
required a time which is indefinite. . . . We find no vestige
of a beginning, no prospect of an end.
                    —James Hutton, *Theory of the Earth*, 1785

Thus, life on earth has often been disturbed by terrible events.
                    —Georges Cuvier, "Preliminary Discourse," 1812

James Hutton, son of a wealthy Scottish landowner, was a born
dabbler—which was exactly what the new science needed.

In 1740, at age fourteen, Hutton went to the University of
Edinburgh to study Greek and Latin. He abandoned the classics
to try his hand at chemistry, left school altogether and apprenticed
himself to a lawyer, gave up the law at twenty-one to study medicine in Edinburgh, wandered off to Paris to specialize in anatomy,
finally finished his medical degree, and then decided that the life
of a doctor wasn't for him. Instead, with the help of a friend, he
started a company that produced the industrial chemical known as
sal ammoniac. The company was successful, but before long Hutton left its management to his friend and decided instead to run
the family farms.

He experimented with agriculture; he joined the Philosophical
Society of Edinburgh and presented papers on botany, mineralogy, and artillery; he carried out chemical experiments; he visited

salt mines, hiked into mountains to examine geologic formations, researched the production of coal. "Free from the interruption of professional avocations," writes his friend and biographer John Playfair, "he enjoyed the entire command of his own time, and had sufficient energy of mind to afford himself continual occupation."[1]

This sort of wide-ranging interest, combined with a private income that made earning a living unnecessary, was a prerequisite for the study of the earth—a field that still had no university chairs, no definition, no borders, no commonly accepted name. Buffon had called himself a "historian"; his fellow investigators of the planet called themselves astronomers, mathematicians, natural philosophers, gentlemen. Hutton himself spent the first half of his life bouncing among the identities of chemist, manufacturer, and farmer. He didn't publish his first study until the age of fifty, a pamphlet with the scintillating title *Considerations on the Nature, Quality, and Distinctions of Coal and Culm.** It was not exactly a best seller ("very ingenious and satisfactory," Playfair assures us, "though, perhaps, considering the purpose for which it was written, it is on too scientific a plan"), but it reveals Hutton's slowly focusing interests. Increasingly, he was studying the composition of the earth itself.[2]

The year after Hutton's pamphlet appeared, the Swiss mathematician Jean André Deluc gave a name to the science of the earth.

For some years, Deluc had been carrying on a correspondence with Queen Charlotte of England and, as socially ambitious thinkers were prone to do, had collected the letters for publication. The first volume (there were a lot of letters), published in 1778, was titled *Physical and Moral Letters on the Mountains and on the History of the Earth and Man.* Deluc was a devout Protestant, preoccupied with reconciling his theories about rock formation and soil strata with the Genesis account (he disagreed strongly with Buffon's chronology, preferring an age of about 10,000 years for the planet). "These letters," Deluc begins, "are only an outline of a Treatise on Cosmology." And then, in a footnote, he laments the inaccuracy of

---

* "Culm" was the term used for a particular kind of anthracite found in the southwest of England.

the term: "I mean here by cosmology only the knowledge of the earth, and not that of the universe. In this sense, 'geology' would have been the correct word, but I dare not adopt it, because it is not in common use."[3]

Despite his disclaimer, Deluc continued to use the term. The *Letters* grew in popularity, and the word "geology" gained currency. The study of the earth still had diffuse borders, mingling with theology and biblical studies on one side, physics and astronomy on another, metallurgy and chemistry on a third; but at last, it had a label.

In 1783, six years after the publication of his pamphlet on coal, Hutton finally assembled his evolving geologic theories into a major public presentation. The Philosophical Society of Edinburgh had just merged with the brand-new Royal Society of Edinburgh, dedicated to the study of both philosophy and natural sciences; anxious to support the new institution, Hutton agreed to give a paper on the "terrestrial system." "The institution of the Royal Society," Playfair explains, "had the good effect of calling forth from Dr. Hutton the first sketch of a theory of the earth, the formation of which had been the great object of his life."[4]

Hutton was now fifty-seven years old. He had become a "skilful mineralogist" and had "examined the great facts of geology with his own eyes"; he was "eminently skilled in physical geography" and in chemistry; he had read widely in natural history. He had put all of these skills to use in investigating the nature of the earth and its systems, but until now, he had shared few of his ideas with others.[5]

On March 7, 1785, the first part of his presentation was read out to the Royal Society by one of his friends; Hutton himself pleaded illness (possibly overcome by nerves). "The first part of his Discourse," the minutes of the Royal Society record, "contain[ed] an Examination of the System of the habitable Earth, with regard to its Duration and Stability." A month later, at the society's April meeting, Hutton himself recapped his March presentation and then read the rest of the paper.

All of the layers of land we now see, Hutton argued, had once been loose materials, washing around at the bottom of oceans; heat had fused these materials together and pushed them up above the ocean's surface, producing dry land. This dry land was, in turn, constantly worn away by the waters around it, in a repeating cycle. This process could still be observed in the oceans and continents of the present day.

In other words, the earth around us had been formed not by huge past catastrophic events, not by extraordinary interventions, but simply through the exact same ebbs and flows, waves and recesses, accumulations and erosions, that still continue.[6]

Whatever happened in the past could be explained simply by observing the present; or, as later thinkers would phrase it, "the present is the key to the past." There was *uniformity* between the natural phenomena we now see and the natural phenomena of epochs past—and, by inference, the natural phenomena of the future. "We are to examine the construction of the present earth," Hutton explained, "in order to understand the natural operations of time past; [and] to acquire principles, by which we may conclude with regard to the future course of things."[7]

Accepting the principle of uniformity (there were no great past catastrophes; the earth has been formed by the natural processes we can still measure) led to another conclusion. Those processes, which produce change very, very slowly, must have been going on for a *very long time*. "To sum up the argument," Hutton wrote, "we are certain that . . . this operation is so extremely slow, that we cannot find a measure of the quantity in order to form an estimate. . . . The production of our present continents must have required a time which is indefinite."[8]

Indefinite: outside of our ability to define. At the end of the first chapter of his paper, Hutton put this in even stronger terms: "The result, therefore, of this physical inquiry," he concluded, "is that we find no vestige of a beginning, no prospect of an end." This was not the Greek version of an eternal and unchanging earth, but rather a vision of a physical world that changes on an entirely different timescale than the one we inhabit. Hutton's paper put into words what Buffon had only implied: that geologic time, "deep time," is so different from the time of human experience that we can barely even use the measure of years to express it.[9]

This was even more transgressive of the Genesis story than Buffon's relatively mild epochs had been, which may account for Hutton's attack of nervous vapors on his first scheduled appearance before the Royal Society. Oddly, though, his paper—published three years later under the title *Theory of the Earth*—attracted relatively little ire. His biographer Playfair blames intellectual exhaustion: "The world was tired out with unsuccessful attempts to form geological theories," he complains, "by men often but ill informed of the phenomena which they proposed to explain."[10]

Jean André Deluc published a rebuttal of Hutton's chronology, as did the Irish chemist Richard Kirwan, who accused Hutton of theorizing "contrary to reason and the tenor of the Mosaic history." But geology was still a brand-new field, its practitioners were widely dispersed, and ten years after the publication of *Theory of the Earth*, reaction to Hutton's theories of uniformity and deep time were still scattered. "This Theory of the Earth," Playfair complains, "[should] have produced a sudden and visible effect, and . . . men of science would have been everywhere eager to decide concerning its real value. Yet the truth is, that it drew their attention very slowly."[11]

Part of the blame may lie with Hutton's writing, which tends toward the serpentine. (As Dennis Dean puts it, Hutton was "almost entirely innocent of rhetorical accomplishments.") And Hutton also failed to explain whether *any* sort of catastrophe *could* have taken place in the past—which was confusing, given that sediment deposits and fossil strata seemed to testify to *some* unusual events, *sometime* in the past. But the *Theory of the Earth* was as much philosophy as science; Hutton was attempting to establish a general principle for earth science ("We cannot know anything about the past except through the lens of the present"), not to interpret the history of particular geologic layers.[12]

James Hutton, who was suffering from chronic kidney failure, did not live to see his theory either fully embraced or decisively rejected. He spent the next ten years working, slowly and painfully, on a revision and expansion of *Theory of the Earth*; it was published in 1795 and turned out to be even more obscure than the original.

In March of 1797, after a long day of writing, he was seized with fits of shivering and cramps. He sent for his personal physician but died just as the doctor arrived.[13]

.

One year before Hutton's death, a young French naturalist named Georges Cuvier presented his first major paper to the National Institute of Sciences and Arts in Paris.

Cuvier, twenty-seven years old, had recently been elected to membership in the National Institute (the French equivalent of the Royal Society). He had studied in both France and Germany, mastering both Aristotle's *History of Animals* and Buffon's thirty-five-volume *Natural History*, and he had recently been appointed to a position at the National Museum of Natural History in Paris, where his job was to organize and catalogue a massive collection of fossil bones that had never been properly sorted (a "charnel house," he called it) and also to deliver public lectures on animal anatomy.

The 1796 paper *Mémoires sur les espèces d'éléphants vivants et fossiles* ("On the Species of Living and Fossil Elephants") compared the skeletons of Indian and African elephants to fossilized bones found in Siberia. These bones were thought by many naturalists to belong to ancient elephants, but Cuvier argued that differences in the skull shape, tusks, and teeth proved that the fossil bones belonged to an entirely different species—a "mammoth" animal, not an elephant at all, that had been wiped out. The "mammoth," he concluded, was extinct—a "lost species" that no longer lived on the earth.[14]

This was a highly controversial statement.

After Nicholas Steno, most naturalists had come to accept that fossils were remains of living creatures instead of oddly shaped rock formations. But the idea that whole species of animals might have died out, in the past, was problematic from three different directions. It posed a theological difficulty: how could God have created animals that weren't designed well enough to survive? It contradicted Aristotelian principles of biology, still accepted by many students of animal anatomy: animals had developed their structure so that they could function well and survive in their environment. And it made no sense in terms of Hutton's recently proposed theory of infinitesimally gradual change: this change was surely too slow to wipe out entire classes of creatures, so recently that their bones still existed.

Many students of animal life continued to insist that fossils (like the Siberian mammoth) were variations on still-existing species, or that (like the mollusk remnants known as ammonites) they still existed in the very deep ocean, or somewhere that couldn't be easily investigated by humans. But Cuvier disagreed.

He had, so far, no explanation for *why* some species had died out. He had not started out with a grand, overarching theory of life and tried to fit the mammoth fossils into it. Instead, he had exercised the scientific, Baconian method: close and careful examination of specific natural phenomena. This examination had led him to a conclusion: The mammoth was not an elephant. It was something else, and it no longer existed.

As he attempted to put together a history of different animal species that had populated (or still roamed) the earth, Cuvier found himself, unexpectedly, constructing a history of the earth itself. "There is a science that does not appear at first sight to have such close affinities with anatomy," he noted in his 1796 paper,

> one that is concerned with the structure of the earth, that collects the monuments of the physical history of the globe, and tries with a bold hand to sketch a picture of the revolutions it has undergone; in a word, it is only with the help of anatomy that geology can establish in a sure manner several of the facts that can serve as its foundations.[15]

Revolutions: hardly an innocuous word in a country that had seen the storming of the Bastille only seven years earlier. Already germinating in Cuvier's mind was a possible explanation—that the mammoth, not to mention other fossil species (such as an enormous skeleton found in Ohio, which he later labeled the "mastodon"), might have been wiped out by an extraordinary, globe-changing, onetime catastrophe.

To Cuvier, nothing seemed more likely than that the history of the globe might mirror the stormy transitions of human soci-

ety. And so, at the very end of his careful analysis of skeletons, he
detoured into speculation:

> What has become of these two enormous animals of which one
> no longer finds any traces, and so many others of which the
> remains are found everywhere on earth, and of which perhaps
> none still exist? . . . All of these facts . . . seem to me to prove the
> existence of a world previous to ours, destroyed by some kind of
> catastrophe. But what was this primitive earth? What was this
> nature that was not subject to man's dominion? And what revolu-
> tion was able to wipe it out, to the point of leaving no trace of it
> except some half-decomposed bones?

Committed to Baconian science, Cuvier could merely pose these
questions, not answer them with any kind of certainty:

> It is not for us to involve ourselves in the vast field of conjectures
> that these questions open up. Only more daring philosophers
> undertake that. Modest anatomy, restricted to detailed study and
> to the scrupulous comparison of the objects submitted to its eyes
> and its scalpel, will be content with the honor of having opened
> up this new highway to the geniuses who will dare to follow it.[16]

For the next three or four years, Cuvier concentrated on "mod-
est anatomy," analyzing and categorizing the bones in his "charnel
house." By 1800 he had identified twenty-three new species, all of
which appeared to be extinct. And increasingly, he found himself
pushed toward that "vast field of conjectures"—a theory of the
earth itself. In a series of public lectures given in 1804 and 1805,
he proposed that fossil beds revealed the "strongest proofs that the
globe has not always been as it is at present." The rock layers in
which fossils were found could be used to construct a time line of
the earth's development; the fossil strata were a book of the earth's
past that could be read by the perceptive. Among his suggestions:

> The parts that contain no organized bodies at all are the most
> ancient. Thus [life] has not always existed.

There have been several successive changes in state, from sea into land, from land into sea.

There have been different ages, producing different kinds of fossils.

Several of the revolutions that have changed the state of the globe have been sudden.[17]

Cuvier had moved from "conjecture" to hypothesis: the book of the earth, he now believed, was peppered with catastrophes.

While preparing and delivering his public lectures, he had also been working with one of his colleagues at the National Museum, the mineralogist Alexandre Brongniart, on a project analyzing the rock strata around Paris. The city sits in a 7,000-square-mile basin of sedimentary rock; Cuvier and Brongniart had been carefully constructing a cross-sectional map of the basin's layers. In 1808 they presented their findings to the National Institute, and in 1811 the findings were printed, in an expanded form, for the public to read.

The *bassin de Paris*, they explained, was made up of an ancient foundation of chalk, over which successive layers had been deposited, one by one. Each of these layers contained unique fossils. Including the chalk, there were six distinct layers in the Paris Basin; six different eras in the earth's past, each with its own population of plants and animals, some now extinct.

This orderly reading of a sequential past into confused and often disturbed layers caused a minor sensation in the world of natural philosophy; in both Europe and Britain, other mineralogists and "geologists" (still a brand-new term) set themselves to analyzing local layers of rock in the same way. Cuvier himself took an even wider view. The six layers of the Paris Basin, he soon concluded, were a microcosm of the planet; and he was quick to extrapolate his discoveries into an earth-wide theory.

He published this theory in 1812 as the first section of his collected papers on fossils (*Recherches sur les ossemens fossiles de quadrupèdes*, an anthology of all the different studies he had presented and published since 1804). The *Recherches* was a technical work for specialists, but the first section, which Cuvier titled "Preliminary Discourse," was intended for the general public.[18]

The earth, Cuvier argued, had undergone six separate catastrophic changes. The fossil content of each Paris layer changed suddenly and distinctly, not gradually and by degrees (as Hutton's theory of uniformity suggested); therefore, it seemed clear that a series of nearly worldwide disasters had wiped out various populations of flora and fauna, leaving only small unaffected areas from which surviving animals and plants could then migrate over the newly changed surface of the earth. The "Discourse" covers, carefully and methodically, the proofs of these six calamities in fossil beds, in soil strata, and in mountain rocks. "Thus, life on earth has often been disturbed by terrible events," Cuvier concluded, "calamities which initially perhaps shook the entire crust of the earth to a great depth. . . . These great and terrible events are clearly imprinted everywhere, for the eye that knows how to read." Cuvier was working on the same methodological path as Hutton, but the close examination of the earth had led the two men in opposite directions: one toward an earth history that rumbled along in a steady, unspectacular, single direction; the other, toward a past punctuated by multiple unexpected disasters.[19]

The "Preliminary Discourse" was translated and republished separately again and again, reaching an even wider audience than Cuvier had hoped. The theory of repeating catastrophes struck a chord in his readers—not just the French citizens who were living in the wake of the recent revolution, but also the British and European readers who had been raised with the biblical accounts. Cuvier's six "eras" were easily reconciled with the six days of creation; his extraordinary events recollected creation *ex nihilo*, the Fall, the Great Deluge.

Which was, in all likelihood, not Cuvier's intention. He was reading the layers of the Paris Basin, not the book of Genesis. At the very end of the "Discourse" he suggests that the biblical Flood, as well as flood accounts from China and India, might preserve the memory of the last of his six catastrophes: "All known traditions make the renewal of society reach back to a major catastrophe," he writes, "[but] the date cannot reach back much more than five or six thousand years. . . . But these countries . . . had already been inhabited previously, if not by men, then at least by terrestrial animals." He had come to his conclusion because of the physical evidence, not the biblical story.

In fact, although the Great Deluge might mark the beginning of human civilization, the era of human existence is only a small and recent one, compared with the entire history of the earth: "Man," Cuvier concludes, "to whom has been accorded only an instant on earth, [may now have] the glory of reconstructing the history of the thousands of centuries that preceded his existence, and of the thousands of beings that have not been his contemporaries."[20]

This, too, was a version of deep time, of thousands of centuries preceding the human race.

Cuvier's catastrophism proposed a very different mechanism for change than Hutton's long, slow uniformity of process. Over the next decades, uniformity and catastrophism would struggle for control over the interpretation of the earth's raw data. But despite their differences, the two theories shared a common commitment: both would find Bishop Ussher's date far, far too recent to make sense of the history of the earth.

To read relevant excerpts from the Theory of the Earth, *visit http:// susanwisebauer.com/story-of-science.*

JAMES HUTTON
Theory of the Earth
(1785/1788)

James Hutton's prose style does him no favors; even his biographer Playfair, one of his greatest fans, remarks delicately that the reasoning is "sometimes embarrassed by the care taken to render it strictly logical," and that the "transitions, from the author's peculiar notions of arrangement, are often unexpected and abrupt." But Hutton's argument for a long and uniform history provided the foundation for modern geology. It isn't necessary to plow through the entire, unnecessarily obscure *Theory of the Earth*, but be sure to read the first chapter, which outlines the basic method by which

Hutton believes the continents took shape, and also presents Hutton's version of deep time.

Although there are several versions of the original 1785 and 1788 English text archived online, they use unmodernized spelling, and the scans are often indistinct. A digital version with modern typesetting is available for Kindle.

James Hutton, *Theory of the Earth*, Amazon Digital Services (e-book, no date, ASIN B0071FII7O).

A paperback reprint is available from Kessinger Publishing.

James Hutton, *Theory of the Earth*, Kessinger Publishing (paperback, 2010, ISBN 978-1162713540).

To read relevant excerpts from "Preliminary Discourse," visit http://susan wisebauer.com/story-of-science.

GEORGES CUVIER
"Preliminary Discourse"
(1812)

The best modern translation of Cuvier is found in an anthology of Cuvier's works collected and translated by Martin J. S. Rudwick. It contains not only the "Preliminary Discourse," but excerpts from Cuvier's groundbreaking 1796 paper, the 1811 publication on the Paris Basin, and other works.

The "Preliminary Discourse," or "The Revolutions of the Globe" (the title often given the discourse when published separately), is found in Chapter 15. There is no need to read all of Rudwick's preface, which is almost as long as the "Discourse" but less elegantly written.

Martin J. S. Rudwick, *Georges Cuvier, Fossil Bones, and Geological Catastrophes: New Translations & Interpretations of the Primary Texts*, University of Chicago Press (paperback and e-book, 1998, ISBN 978-0226731070).

Robert Jameson's 1818 translation, published under the title *Essay on the Theory of the Earth*, is archaic in spots but still perfectly readable; it can be found in a number of free e-book versions, as well as in many library collections.

Georges Cuvier, *Essay on the Theory of the Earth*, trans. Robert Jameson, Kirk & Mercein (hardcover [out of print] and e-book, 1818, no ISBN).

FIFTEEN

A Long and Steady History

Uniformitarianism becomes the norm

> The order of nature has, from the earliest periods, been uniform.
>
> —Charles Lyell, *Principles of Geology*, 1830

In 1830, the geologist Charles Lyell weighed in on Hutton's side of the argument. And his arguments for long, slow change were so forceful that catastrophism was thrown out of the geologists' vocabulary—for a century and a half.

Charles Lyell, a fellow Scot, had been born the same year that Hutton died. He had gone up to Oxford in 1816, intending to read law; but the earth was his hobby, and he spent a good part of his second year attending the geology lectures given at Corpus Christi by William Buckland.

Buckland, himself trained in chemistry and mineralogy, was an ordained clergyman as well as an enthusiastic geologist. He was a disciple of Cuvier, a vigorous supporter of catastrophism, and (like the majority of his contemporaries) a believer in Genesis. Like the fictional Telliamed, he saw no conflict between geology and faith: "As far as it goes," he explained in his lectures, "the Mosaic account is in perfect harmony with the discoveries of modern science." The creation account in Genesis hit the high points of the earth's history (its original creation and the rise of the human race); geology filled in the details.

If Geology goes further, and shews that the present system of this planet is built on the wreck and ruins of one more ancient, there is nothing in this inconsistent with the Mosaic declaration, that the whole material universe was created in the beginning by the Almighty: and though Moses confines the detail of his history to the preparation of this globe for the reception of the human race, he does not deny the prior existence of another system of things.[1]

Young Lyell found this pronouncement entirely persuasive. In 1818 he accompanied Buckland on a field trip to Paris: "Went to the Jardin des Plantes," he wrote in his diary, "where I again looked over Cuvier's lecture-room, filled with fossil remains, among which are three glorious relics of a former world." Those former worlds had been wrecked multiple times, in the gap between God's original creation of the earth and the divine reintervention that brought about man.[2]

But as time went on, Lyell found himself less and less satisfied by Cuvier's scheme. He had plenty of time to think about it, since he was (like most early geologists) independently wealthy; after Oxford he made a desultory stab at practicing the law and then gave it up to spend his time mastering all of the different fields of knowledge necessary for the study of the earth (his own list included "chemistry, natural philosophy, mineralogy, zoology, comparative anatomy, botany; in short, . . . every science relating to organic and inorganic nature"). He spent time in the field, journeying throughout Scotland to examine the sediment layers in rocky hillsides and remote lakes; he joined the Geological Society of London and presented papers about his findings; he corresponded with other earth scientists.

In 1825 Lyell made the tentative suggestion that catastrophe wasn't *necessarily* the cause of past phenomena. "In the present state of our knowledge," he wrote in the London journal *Quarterly Review*, "it appears premature to assume that existing agents could not, in the lapse of ages, produce such effects." Extraordinary, earth-wrecking disasters *could* have produced the specimens in Cuvier's collections. It had certainly not been *disproven*, though, that the "existing agents" still at work in the world—plain old erosion, the common rise and fall of temperatures, the regular wash of the tides—might be responsible instead.[3]

It would just take them a whole lot longer.

For several more years, Lyell looked for proof that those "existing agents" could, over time, act with as much power as Cuvier's catastrophes. Three years later, in the Limagne plain of France, he found tracks of an ancient riverbed that seemed to have eaten an enormous trench into granite and lava layers, in a pattern that could not possibly be chalked up to deluge or upheaval, but only to the long, slow "progress of ages." "This is an astonishing proof," Lyell wrote back to his father, "of what a river can do in some thousand or hundred thousand years by its continual wearing." Afterward he traveled across Sicily, ascended Mount Etna, and hiked into the south of Italy, all the time finding more and more geologic evidence that ancient and modern causes were the same.[4]

Underlying all of these discoveries was Lyell's growing conviction that catastrophism was a dead end for geology as a science. If onetime past events were responsible for the current form of the earth, there was no way that the geologist could truly understand the present by exercising reason. The geologist could always haul in a disastrous flood, or a passing giant comet, or some other event that could never be experimentally reproduced, to explain the planet. Geology would remain the study of *history*, mixing story and interpretation with observation, filled with speculations that could never be scientifically *proved*.

By the end of his Italian trip in 1829, Lyell had determined to lay out the principles that would make geology into a true *science*. Sitting in a Naples inn, he wrote to a friend that his book on the subject was already "in part written and all planned":

> It will not pretend to give even an abstract of all that is known in geology, but it will endeavour to establish the *principle of reasoning* in the science; and all my geology will come in as illustration of my views of those principles.

Those principles would make it possible for geologists to conduct their science with the same rigor that Newtonian physics or Galilean astronomy demanded:

> *No causes whatever* have from the earliest time to which we can look back, to the present, ever acted, but those *now acting*; and . . . they never acted with different degrees of energy from that which they now exert.[5]

Only forces that could be *presently observed* would be admitted to Lyell's encyclopedia of explanations. He intended, he wrote in another letter, to understand the earth

> without help from a comet, or any astronomical change, or any cooling down of the original red-hot nucleus, or any change of inclination of axis or central heat, or volcanic hot vapours and waters and other nostrums, but all easily and naturally.[6]

When Lyell published the first volume of his guide to geology in the following year, he chose a title that made this commitment perfectly clear: *Principles of Geology, Being an Attempt to Explain the Former Changes of the Earth's Surface, by Reference to Causes Now in Operation.*

Over the course of twenty-six short chapters, Lyell laid out three interlocking principles for geology, now generally known by the names *actualism, anticatastrophism,* and (more awkwardly) *the earth as a steady-state system.*

> *Actualism.* Every force that has acted in the past is still acting (and can be observed) in the present.

> *Anticatastrophism.* Those forces did not act with more intensity in the past; their *degree* has not changed.

> *The earth as a steady-state system.* The history of the earth has no direction or progression; all periods are essentially the same.[7]

Two years later, the English natural philosopher and clergyman William Whewell gave Lyell's principles the label by which they have been known ever since: *uniformitarianism*.

The *Principles of Geology* instantly sold out. Lyell's careful marshaling of evidence, his clear writing, and his repeated (perhaps not entirely sincere) assurances that he *did* also believe in a Designer won most of his English readers over at once. Over half a century later, the naturalist and philosopher Alfred Russel Wallace summed up the British success of the *Principles*:

> But in 1830, while Cuvier was at the height of his fame and his book was still being translated into foreign languages, a hitherto unknown writer published the first volume of a work which struck at the very root of the catastrophic theory, and demonstrated, by a vast array of facts and the most cogent reasoning, that almost every portion of it was more or less imaginary and in opposition to the plainest teachings of nature. The victory was complete. From the date of the publication of the "Principles of Geology" there were no more English editions of "The Theory of the Earth."[8]

The theory took a little longer to catch on in Europe. But it was so clear, so rational, so *sensible*, that geologists across the globe ultimately embraced it.

And continued to embrace it. In the 1960s, writes American geologist Walter Alvarez, uniformitarianism was *still* the "dogma" of the field, and "the established wisdom of our science was that nothing really dramatic—no catastrophes—had ever happened in the planet's past." Long, slow, gradual change had become the *only* change admissible in the new science; Lyell had triumphed.[9]

To read relevant excerpts from the Principles of Geology, *visit http:// susanwisebauer.com/story-of-science.*

CHARLES LYELL
Principles of Geology
(1830–32)

Most available editions of the *Principles of Geology* contain all three volumes, written between 1830 and 1832. Originally, Lyell had planned to write just two volumes, one dealing with his overall principles (Volume 1), and the second providing more specific geologic proofs (now Volume 3). Eventually, though, he realized that he had to give some accounting for the fossil record, so he interposed a new volume (the current Volume 2). For the purposes of geology, you need to read only Volume 1, which lays out the principles of uniformitarianism.

The original 1830 text can be read online or downloaded as a PDF from multiple sources. Penguin has also produced a high-quality paperback, edited by James A. Secord, containing a useful introduction and all three volumes.

Sir Charles Lyell, *Principles of Geology, Being an Attempt to Explain the Former Changes of the Earth's Surface, by Reference to Causes Now in Operation*, vol. 1, John Murray (e-book, 1930, no ISBN).

Charles Lyell, *Principles of Geology*, ed. James A. Secord, Penguin Books (paperback and e-book, 1997, ISBN 978-0140435283).

SIXTEEN

The Unanswered Question

Calculating the age of the earth

Many of the fundamental problems of geology can be solved only with reference to the processes involved in the making of the earth.

—Arthur Holmes, *The Age of the Earth*, 1913

Lyell's fierce campaign against extraordinary events had been astoundingly successful.

In fact, nearly a hundred years after the publication of the *Principles*, it remained quite déclassé for a geologist to theorize about past events in the earth's history that could not be explained by the present. Such speculations smacked of seventeenth-century theologizing, of biblical creationism, of the years before geology became a *science*.

The problem was that strict uniformitarianism left the single biggest question in geology unanswered.

Like the fictional Telliamed, Charles Lyell had tried to evade the question of origins by assigning it to religion, not geology. "Probably there was a beginning," he wrote to an early reviewer of the *Principles*, "—it is a metaphysical question, worthy a theologian— probably there will be an end." But that was as far as he could go. Uniformitarianism made it impossible for him to contemplate *any* theory of origins, because beginnings (and endings) implied that the past (and future) *did* differ from the present.

Lyell could not entertain the idea that the earth had originally been a molten ball cooling over an incredibly long time, any more than he could agree that God had opened the heavens and sent a miraculous deluge in the days of Noah. Theories of beginnings had the potential to short-circuit Baconian reasoning, to introduce new and inexplicable past causes as an easy alternative to true understanding. "All I ask," Lyell wrote, just before publication of the *Principles*, "is that at any given period of the past, don't stop inquiry when puzzled by refuge to a 'beginning.' . . . We are called upon to say in each case, 'Which is now most probable, my ignorance of all possible effects of existing causes,' or that 'the beginning' is the cause of this puzzling phenomenon?"[1]

But the earth was a complicated object of study. Unlike the laws of physics, or the principles of chemistry, the earth was a *place*, a thing with a history. It bore on its surface the tracks of a long past. It was home to a species that lived in *time*. "John Heddon, aged 31," began a typical nineteenth-century English news report, "and Thomas Gaydon, aged 68, fell over a scaffolding to the ground." The age of the men involved might seem to have nothing to do with the event; but the time that someone (or something) has been in existence changes the way that human beings understand it. We do not easily exist in ahistoricity.[2]

So it is not surprising that, post-Lyell, a number of thinkers attempted to reconcile the general principles of uniformity with the human compulsion to estimate the age of the earth.

Thirty years after the *Principles*, the Belfast-born mathematician William Thomson—later known as Lord Kelvin—applied a universal law to the solar system and came up with both an end and a possible beginning. The universal law was the Second Law of Thermodynamics: When energy is converted from one form into another, some of it is always expended in the process.* The sun is a natural engine that converts energy into heat. Ergo, the sun is losing energy with every conversion. So in the past, it must have been hotter than it now is; and in the future, it will continue to cool.

*The Second Law of Thermodynamics is, perhaps, better known in its related but simpler phrasing: The universe tends toward entropy. (The gradual, continuing loss of energy is the reason that natural engines in the universe will, ultimately, stop functioning.)

In other words, uniformity didn't imply *stasis*, lack of development. If we accept, Thomson wrote, that "known operations going on at present in the material world" are the only ones that have acted in the past, the conclusion is simple: "Within a finite period of time the earth must have been, and within a finite period of time to come the earth must again be, unfit for the habitation of man." Uniformity meant that the earth had *changed*, drastically, over time. The sun itself could not be older than five hundred million years (with a hundred million years a more likely figure); so the earth could not have orbited it eternally, and one day the earth would cease to orbit it again.[3]

Two years later the Irish physicist John Joly used the amount of sodium that had leached from the earth into ocean water (an ongoing, observable process) to estimate that the oceans were about a hundred million years old. His effort, like Thomson's, honored Lyell's principles.

One hundred million years was a long time. Five hundred million years was even longer. But neither was quite long enough to accommodate Lyell's strict uniformitarianism. Those slow, gradual transformations needed even more *time*.

•

In the first quarter of the twentieth century, a new method of dating arrived on the geologic horizon.

In 1895 the German physicist Wilhelm Roentgen had observed "mysterious rays," streams of energy particles that could penetrate solid matter, in his laboratory; he called these "X-rays," not having any other way to label them. The following year, another physicist, the Parisian Antoine-Henri Becquerel, discovered similar "rays" emitted from uranium salts. In 1898, Marie S. Curie and Pierre Curie observed the same phenomenon, this time rising from thorium. The Curies named the rays *radioactivity* and suggested that the particles streaming from thorium and other elements came not from molecules, but from individual atoms themselves.*

---

*The formulation of atomic theory is covered, briefly, in Chapter 26.

Four years later, physicist Ernest Rutherford and chemist Frederick Soddy concluded that those atoms were actually *disintegrating*. They were unstable (perhaps because they were too massive, or had too much energy) and so were throwing off particles to try to keep their equilibrium. And the rate of their disintegration was measurable, constant, predictable.[4]

Which meant, Ernest Rutherford suggested, that minerals containing an unstable element could be dated by measuring the progress of the decay: "a most valuable method of computing their age," he wrote, in 1906. "Indeed, it appears probable that it will prove one of the most reliable methods of determining the age of the various geological formations." He was unable to give a precise measurement, but he had a range in mind: "Many of the primary radioactive minerals," he concluded, "were undoubtedly deposited at the surface of the earth 100 to 1000 million years ago."[5]

One hundred million years was no shock to Rutherford's readers. But one *thousand* million years—that was moving toward an entirely different time frame.

•

In 1906, Arthur Holmes was sixteen years old: raised as a devout Methodist, preparing to study physics at university, perplexed by the marginal notes in his Bible. "I can still remember the magic fascination of the date of Creation, 4004 BC, which then appeared in the margin of the first page of the Book of Genesis," he noted, later in his life. "I was puzzled by the odd '4.' Why not a nice round 4000 years? And why such a recent date? And how could anyone know?" His physics studies at the Royal College of Science coincided with the rising interest in Rutherford's new dating techniques; halfway through his degree, Arthur Holmes changed his field from physics to geology.[6]

Geology was in a mild uproar. William Thomson (who had become Baron Kelvin in 1892) had just died, at eighty-three, leaving behind an estimate of the earth's age that had steadily shrunk from his original upper estimate of five hundred million years down closer to twenty million as he had calculated and recalculated. Rutherford's "radioactive dating" had yielded wildly varying results; the breakdown of radioactive elements was still poorly understood, and measuring the decay left behind was a complicated and variable business. Arthur Holmes, first as an undergraduate and then as a doctoral student, immersed himself in the brand-new and entirely unstable field of radioactivity.

He was convinced that radioactive dating could yield much more precise results: "Every radioactive mineral," he wrote, immediately after receiving his doctorate, "can be regarded as a chronometer registering its own age with *exquisite accuracy*."[7]

Exquisite accuracy still lay in the future, though. In one of Holmes's earliest projects, he made use of the decay rates introduced by Rutherford and Soddy to date a Norwegian rock layer at 370 million years. Shortly afterward, he calculated another rock sample at 1.6 billion years. And when, at the age of twenty-three, he wrote his first book, a good part of it was devoted to explaining *why* radioactive methods produced such a wild rangeof dates.[8]

The Age of the Earth, published in 1913, still stands as a benchmark in geology—but not because it answered the implied question. (In fact, Holmes never actually specifies in it what the age of the earth might be; that would come much later in his life.)

Rather, *The Age of the Earth* shifted the trajectory of the adolescent science slightly, angling it away from Lyell's strict uniformitarianism.

Uniformity itself Holmes had no problem with. In fact, he thought that both Hutton and Lyell had done geology a favor by forcing it to give up its theories of origins. "Speculative fancies concerning the origin of the world," he wrote,

> form the subject-matter of many of the earliest writings on record, and throughout the intellectual history of mankind the problem has proved to be one of supreme fascination. It was not, however, until quite recent times that the efforts of imagination gave place to reasoned hypotheses. . . . At first, on having attained the status of a science, geology steadfastly refused to consider seriously the cosmogonic fantasies then current. It was Hutton, who by advocating the direct observation of nature in place of

the old scholastic arguments, first delivered geology from the inevitable wranglings that would necessarily have arisen from so premature a discussion of the beginning of things.

But the question could not be ignored forever:

> While today we are still unable from geological facts alone to trace back with confidence the details of the earth's beginning, yet the uncertainty which justified Hutton in entirely disregarding the earth's genesis no longer exists. Astronomy, physics, and chemistry have all . . . done much to remove our modern ideas from the dangerous quicksands of speculation.[9]

Lord Kelvin, Holmes concluded, had done geology a great favor by reintroducing the problem of the earth's age: "He invaded the domain of Geology . . . determined to protest against what he considered the immoderate application of the principle of Uniformity."

But radioactive theory, not the Second Law of Thermodynamics, would unlock the secret of the earth's age. Radioactive minerals, Holmes concluded, are "clocks wound up at the time of their origin," and "we are now confident that the means of reading these time-keepers is in our possession."

The confidence was a tiny bit premature, since radioactive dating was currently yielding ages ranging from 370 million to 1,200 million years. The factors involved were complicated; much was still unknown, and much work remained to be done. Yet Holmes remained confident. "The problem has advanced from the qualitative to the quantitative stage," he concluded, "and for the first time in historical geology, accurate measurement founded on delicate experimental work has become possible."[10]

In 1913, this was more prophecy than statement of fact. But in the next decades, study of radioactive decay yielded more and better data, and estimates of the earth's age began to tighten to a very clear range. In later publications, Holmes moved from a stance of 3,000 million years, to 1.6 billion years, to over 3 billion years; by 1930, Ernest Rutherford had followed him, calculating that the earth had been formed "about 4×10^9 years ago": 4 billion years.[11]

Within thirty years, almost the entire scientific community had shifted its heading onto Holmes's new track.

"Uniformity proved a great advance," Holmes had written, "but in detail it is apt to lead us astray if applied too dogmatically." Uniformity remained the basic assumption of geology; but it was now tempered by the necessity of understanding the earth's history as encompassing a beginning, a direction, and (in all likelihood) an end.[12]

To read relevant excerpts from The Age of the Earth, *visit http://susan wisebauer.com/story-of-science.*

ARTHUR HOLMES
The Age of the Earth
(1913)

The original text of *The Age of the Earth* can be viewed online (see http://susanwisebauer.com/story-of-science for a live link) and is also available as a PDF download from Forgotten Books (readable, but the conversion to a downloadable format has altered spacing and reproduced headings so mechanically that they often appear in the middle of a page). A print copy of the original Harper's Library edition, containing Holmes's illustrations and diagrams, can be found at most university and many large public libraries. Be careful if ordering a physical copy secondhand, since many copies offered are low-quality print-on-demand reproductions of the electronic text. Look instead for the reprints by Ernest Benn and Nelson & Sons.

Arthur Holmes, *The Age of the Earth*, Harper & Brothers (hardcover and e-book, 1913, no ISBN).

SEVENTEEN

The Return of the Grand Theory

Continental drift

> To determine "truth" . . . [is] to find the picture that sets
> out all the known facts in the best arrangementand that
> therefore has the highest degree of probability.
> —Alfred Wegener, *The Origin of Continents and Oceans*, 1915

The decay of radioactive elements had reintroduced, into Lyell's strict uniformitarianism, the ticking clock.

It now seemed impossible to deny that there *had* been a beginning to earth time (even though no one knew exactly when it might be). So Lyell's long, slow, predictable, mostly uniform series of changes had a *start*.

Which led to a series of additional questions. What did the earth look like at that start? How had those slow changes altered its original form? And (in line with Lyell's principles), could those slow changes still be detected, creeping along in the present?

The Austrian geologist Eduard Suess garnered a wide following with his theory of "thermal contraction," which was a throwback to Newton's molten earth cooling over time, combined with Lyellian cycles of erosion and buildup. According to contraction theory, as the newborn, extremely hot earth began to cool into a solid globe, its crust contracted and wrinkled up, like a drying apple. Some of the crust collapsed inward, forming ocean basins and pushing the crust between the collapsed basins upward into continents. Then, as the earth went *on* cooling (over those millions and millions of years that the measurement of radioactive elements now suggested), it contracted further, and the continents collapsed inward and pushed the ocean basins up. This was a repeating cycle, a "continual interchange of land and sea" that pushed fossil remains from the tops of mountains to the bottoms of seas and back up again—which explained why marine fossils now existed on mountains, and why the same kinds of fossil remains could be found on different and disconnected continents.[1]

Other geologists accepted the basic model of contraction but argued that the continents and oceans had always been in the same place ("permanence theory"). Erosion and silting, or perhaps volcanic activity followed by earthquakes, created and destroyed land bridges between them. Animals, and eventually people, wandered across the land bridges, which then disintegrated.

But there were massive problems with contraction theory.

For one thing, the idea of "land bridges popping up and down" (in Naomi Oreskes's phrase) seemed more than a little ad hoc. Radiation theory introduced another problem: it now appeared that certain atoms generated *more* heat over time, which didn't fit at all with the idea that a uniformly hot earth was now cooling.[2]

And calculations made by more than one physicist suggested that the earth was simply too dense for contraction to produce high mountains and cavernous ocean basins. Physicist Osmand Fisher, as far back as 1881, had suggested that perhaps the globe *wasn't* solid; perhaps there was a "fluid substrate" to the earth, a soft and yielding layer farther down that allowed the surface above it to shift, break, and move. Fisher's contemporary, the American geologist C. E. Dutton, had pointed to the movement of glaciers as an example of how a solid layer could "flow" up and down. But there was no way to prove that a fluid underlayer existed, and no proof that the dry and solid surface of the earth had *ever* acted like a sheet of ice.[3]

Alfred Wegener, a German astronomer with a particular interest in the weather, had a different solution.

"Anyone who compares, on a globe, the opposite coasts of South America and Africa," he wrote in 1915, "cannot fail to be struck by the similar configuration of the two coast lines." The jigsaw match suggested to him that the continents had once been a single mass, a

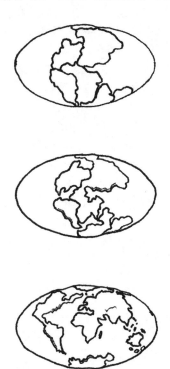

17.1 PANGEA AND CONTINENTAL DRIFT

giant supercontinent that he labeled Pangea; long, long ago, Pangea had broken up and drifted apart.[4]

This theory required him to provide an explanation for how solid earth could "drift." So he proposed that the earth was not actually solid. Instead, it consisted of a liquid core, surrounded by a series of shells that increased in density as they got closer to the surface.

It was a simple and elegant explanation and accounted for almost all the factors that puzzled geologists: the odd similarities between far-distant fossils, the apparent interlocking fit of the continental coastlines, the existence of mountains (which sprang up where the drifting pieces collided and overlapped). And it was greeted, in the world of geology, by shrieks of derision.

The reaction was not entirely irrational. Although continental drift made sense of the map, Wegener had formulated it in the absence of pretty much any other physical evidence. It was a grand theory in the Aristotelian tradition; he had come up with the huge overarching explanation first, and defended it entirely on its internal consistency. This was not particularly "scientific," as the American geologist Harry Fielding Reid sniped in a review: "Science has developed by the painstaking comparison of observations," Reid wrote, "and, through close induction, by taking one short step backward to their cause; not by first guessing at the cause and then deducing the phenomena." The paleontologist Charles Schuchert complained, "The whole trouble in Wegener's hypothesis and in his methods is . . . that he generalizes too easily from other generalizations." Schuchert's colleague Edward Berry, teaching at Johns Hopkins, agreed: "My principal objection to the Wegener hypothesis rests on the author's method," Berry wrote:

> This, in my opinion, is not scientific, but takes the familiar course of an initial idea, a selective search through the literature for corroborative evidence, ignoring most of the facts that are opposed to the idea, and ending in a state of auto-intoxication in which the subjective idea comes to be considered as an objective fact.[5]

On the other hand, there wasn't a great deal of painstaking observation or of concrete fact to support contraction theory either, not to mention vanishing land bridges.

The wholesale resistance to Wegener's intuitive leap may have had something to do with border protection; Wegener was neither a geologist nor a paleontologist. He was a tinkerer in meteorology, an adventurer who had once been forced to eat his own ponies to survive in an icy Greenland camp, a German who had fought on the side of the Central Powers in World War I. But there *were* undeniable weaknesses in drift theory. Wegener was unable to explain the mechanism behind it; he could offer no reason why Pangea didn't simply remain one supercontinent. And continental drift was horribly counterintuitive. It was almost impossible to imagine those enormous landmasses taking to the high seas, plowing through the ocean as if they were the Staten Island Ferry. It required a conceptual leap not unlike the one demanded four hundred years earlier, when the apparently stationary earth was sent hurtling through space at top speeds.

Wegener himself was conscious of the problems, but he believed that the explanatory power of his theory trumped his lack of explicit proof. He argued that, after all, the earth "supplies no direct information" about its configuration:

> We are like a judge confronted by a defendant who declines to answer, and we must determine the truth from the circumstantial evidence. All the proofs we can muster have the deceptive character of this type of evidence. . . . It is only by combining the information furnished by all the earth sciences that we can hope to determine "truth" here, that is to say, to find the picture that sets out all the known facts in the best arrangement and that therefore has the highest degree of probability.[6]

Wegener continued to work on his theory, adding new arguments, reworking and republishing *The Origin of Continents and Oceans*. The book went into a second edition, and then a third. He summarized his arguments for continental drift in lectures and symposiums, in Europe and in North America. "The theory offers solutions for . . . many apparently insoluble problems," he wrote, in 1922.[7]

Most geologists still disagreed. But in 1928 the naval astronomers F. B. Littell and J. C. Hammond compared the longitudes of Washington and Paris in 1913 and in 1927. Their readings revealed beyond a doubt that the distance between the two cities had increased by 4.35 meters—a creep of 0.32 meter per year.

Given that Paris is some 6,000 kilometers from Washington, it would have taken over eighteen million years for the two cities to move that far apart. But the drift was measurable, beyond a doubt. When Wegener published the fourth and final edition of *The Origin of Continents and Oceans* in 1929, he added Littell and Hammond's measurements to his final appendix: "The direction and amount of this change," he concluded, "agree very well with the deductions on the basis of drift theory."[8]

Wegener did not live to see the reaction. While the fourth

edition was coming off the press, he was organizing his fourth expedition to Greenland. He arrived in the spring of 1930 and managed to establish an observation camp at a site called Eismitte, nearly in the center of the island. But everything that could go wrong did: ice was thicker than expected, the weather was unsettled and harsh, his hired team lobbied for more money, supplies failed to arrive, dogsled teams vanished. "The whole business is a big catastrophe," he wrote back to a colleague in late August.[9]

In early November, short on food and facing temperatures plunging far below zero, Wegener and a companion abandoned the observation camp and set out for the better-supplied base camp of Scheideck, 250 miles away. They never arrived. In the spring of 1931, Wegener's body was located in a grave halfway between Eismitte and Scheideck; his colleague had vanished entirely.

•

Without its defender, the theory of continental drift could easily have faded; but its explanatory power was too great.

Arthur Holmes found it particularly appealing, and suggested that the movement of continents might be caused by *convection*— the slow movement of the mantle, the heated layer of the earth beneath the crust. This layer, he theorized, might be fluid enough to cycle in lazy currents, like a simmering stew, moving the crust above it as it turned over.

This turned out to be the right explanation, but Holmes, like Wegener, had no ability to peer beneath the crust for proof. Geology needed an extension of the senses through artificial means, instruments to do for geologists what microscopes and "furnaces" had done for chemistry.

Those instruments finally evolved out of war.

Sonar technology, first developed to allow ships to search for attacking submarines, was redeployed in the 1950s to map the ocean floor. For the first time, geologists could see the features of the deep: continental shelves, abyssal plains, midocean ridges and trenches, an entire underwater terrain. One of the mappers was Harry Hess, who had taught geology at Princeton before the

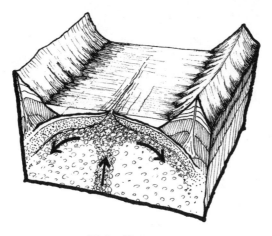

17.2 CONVECTION

war, and commanded a transport outfitted with sonar during the years of combat. In 1962 he published a paper proposing that the newly mapped features of the ocean floor were proof of convection; the midocean ridges were places where the currents, cycling slowly through the mantle, were pushing hot material up into the seafloor, where it formed new crust. Trenches were places where the crust was subsiding back down into the mantle, melting and re-joining the currents. Holmes had been right: "The continents do not plow through oceanic crust impelled by unknown forces," Hess wrote. "Rather, they ride passively on mantle material as it comes to the surface at the crest of the ridge and then moves laterally away from it."[10]

Hess (along with two later Cambridge researchers, F. J. Vine and Drummond Matthews) had provided the foundation for plate tectonics—the theory (finally formulated in the late 1960s) that the earth's crust is made up of separate, continually moving pieces, or *plates*, of crust, that "float" on the mantle of the earth. This, at last, was the mechanism Wegener had been missing. And the grand theory had finally been justified—half a century later.

For a link to Wegener's own brief précis of his argument, written in 1922, visit http://susanwisebauer.com/story-of-science.

ALFRED WEGENER
The Origin of Continents and Oceans
(1915/1929)

John Biram's 1966 translation (made from the fourth German edition of 1929) has been reprinted by Dover Publications and is available in both print and digital formats.

Alfred Wegener, *The Origin of Continents and Oceans*, trans. John Biram, Dover Publications (paperback and e-book, 1966, ISBN 978-0486143897).

EIGHTEEN

Catastrophe, Redux

Bringing extraordinary events back into the earth's history

> After the impact at Chicxulub, 65 million years ago, life on
> Earth was changed forever.
> —Walter Alvarez, *T. rex and the Crater of Doom*, 1997

Just seven years after the first publication of the drift theory, J. Harlen Bretz proposed an even more outrageous explanation for the shape of the earth.

Bretz, a professor of geology at the University of Chicago, had just taken a field party of students to investigate the eastern Washington floodplain known as the Columbia Plateau. It was an eerie basalt landscape, pitted and carved into columns and channels; Bretz nicknamed it the "Channeled Scablands." "Like great scars marring the otherwise fair face to the plateau," he later wrote, "elongated tracts of bare, black rock carved into mazes of buttes and canyons . . . great wounds in the epidermis of soil with which Nature protects the underlying rock." The tortured formations, Bretz thought, could only have been formed by some disastrous event, and the most reasonable explanation was a sudden inundation of water: a long-ago flood, probably caused by a melting glacier, that had suddenly torn down across the plateau, ripping streamways and gouging holes in the basalt plain. In 1923, he published two papers suggesting that the Channeled Scablands had been formed by "a great flood of glacial waters from the north," a "debacle which swept the Columbia Plateau."[1]

Unlike Wegener's theory, this proposal wasn't intended to

account for the form of the entire planet—just one small section of it. But the reaction to it was just as shrill. Immediately, prominent geologists took offense. A 1927 symposium assembled in Washington, DC, was almost entirely devoted to trampling Bretz's theory: the "abnormal flood," as one participant complained, was too "violent an assumption." It was much more reasonable to assume that the Scablands had been shaped by "the same orderly and long-continued process of head-end erosion" that could be observed at any number of rivers and falls.[2]

Bretz had just run, headfirst, into Charles Lyell.

Lyell's drastic uniformitarianism was still being shaped into a more flexible principle, one that would allow the earth to have a beginning, to change significantly over time. But geology remained passionately committed to long, slow change. Sudden catastrophes, even local ones, were considered to be no more "scientific" than those ancient comets and colliding planets.

But Bretz dug in his heels. The features of the Scablands were extraordinary, and they required an extraordinary explanation. And he was willing to buck the academic consensus to provide one.

"Ideas without precedent are generally looked on with disfavor," Bretz wrote in 1928, "and men are shocked if their conceptions of an orderly world are challenged." He spent the next thirty years challenging the uniform "orderliness" of the past: measuring, recording new data, assembling more proofs. In 1956 he published a massive study of the Scablands, demonstrating through the sheer accumulation of observable *facts* that slow and gradual erosion simply could not account for the landscape.[3]

Up to this point, most of Bretz's opponents had not visited the Scablands in person; they had attacked the "abnormal flood" on principle. Slowly, geologists began to file into the Columbia Plateau, study in hand, to see for themselves. Observation only confirmed what Bretz had been insisting all along: a catastrophe had descended on the Scablands. The venerable James Gilluly (much-honored author of *the* geology text used at most universities) stared up at the gashed columns of basalt, shook his head, and said, "Could anyone have been so wrong?" In 1965, a chartered busload of geologists completed a tour of the Scablands by sending

Bretz (now eighty-three years old) a concession by telegram: "We are now," the telegram ended, "all catastrophists."[4]

The slow accumulation of proof had demonstrated the possibility of sudden change. And new information was streaming to the earth from another source: the advancing space program. In 1968 the Apollo 8 expedition to the moon revealed dozens of impact craters; ancient comets (*and* asteroids) had struck not just once, but again and again. This, too, was sudden change—further proof that the sorts of calamities long off limits to geologic explanation *did*, in fact, sometimes happen.

•

In 1968 the American geologist Walter Alvarez was twenty-eight years old and working for a petroleum company in the Netherlands. He had just finished his PhD at Princeton, but the truth was that he was slightly embarrassed by geology. His father was a Nobel Prize–winning physicist, and physics seemed like a much more enthralling field. Physicists were trying to "read the thoughts of God" (as Einstein put it), grappling with relativity, wrestling with quantum mechanics. Meanwhile, geologists were cataloguing rocks, drawing maps, and working for petroleum companies.

But earth science was changing. It had been a discipline focused on a single planet; but now it was "overwhelmed" (as Alvarez himself later wrote) by "data from so many planets and moons that it was hard to remember them all. And most of those bodies were covered with impact craters."[5]

By 1977, Alvarez was teaching at the University of California at Berkeley, and using the new data from space missions to help interpret all those years of mapping and cataloguing rocks. He had discovered an odd phenomenon: a strange abundance of the element iridium in a layer of Italian rock where it had no business being. Radioactive dating put the rock's age at about sixty-five million years—an intriguing date, since it represented the so-called K-T* boundary, a strata of rock where geologists had long noted a dis-

---

*The K-T (Cretaceous-Tertiary) extinction is now more commonly known as the K-Pg (Cretaceous-Paleogene) extinction.

continuity in the fossil record. Before the K-T boundary, dinosaurs and ammonites abounded; after it, they "disappeared forever."[6]

Alvarez consulted his father about the odd iridium layer, and together the two men came up with an outlandish idea. Iridium is much more commonly found in comets and asteroids than in earth layers. Perhaps this iridium was left over from an asteroid collision with the earth. And if it was, could the collision have anything to do with the K-T boundary?

Despite the increasing evidence that asteroids *did* collide with moons and planets, both of these ideas were far out of Alvarez's comfort zone. "In the mid-1970s," Alvarez later wrote, "the thought of a catastrophic event in Earth history was disturbing. As a geology student I had learned that catastrophism is unscientific. I had seen how useful the gradualistic view had been to geologists reading the record of Earth history. I had come to honor it as the doctrine of 'uniformitarianism' and to avoid any mention of catastrophic events in the Earth's past. But Nature seemed to be showing us something quite different." Observation, here as in the Scablands, was quietly disproving uniformitarianism.[7]

Alvarez began to search for iridium layers in the K-T boundary elsewhere on the earth. When they turned up, he proposed, in a 1980 paper published in the journal *Science* (and coauthored by his father, Luis, along with fellow Berkeley scientists Frank Asaro and Helen Michel), that the "KT boundary iridium anomaly" might well be due to an asteroid strike. Furthermore, this impact might explain the fossil discontinuity:

> Impact of a large earth-crossing asteroid would inject about 60 times the object's mass into the atmosphere as pulverized rock; a fraction of this dust would stay in the stratosphere for several years and be distributed worldwide. The resulting darkness would suppress photosynthesis, and the expected biological consequences match quite closely the extinctions observed in the paleontological record.[8]

The theory was greeted with skepticism—but nothing like the scorn and hostility generated by Wegener or Bretz. There was already too much evidence that such a strike was possible. What was unclear was whether it had actually *happened.*

So the skepticism of the 1980s took the form of a vigorous hunt for proof. Since the iridium anomaly (as Alvarez later observed) was "clearly real and probably global, the impact hypothesis attracted hundreds of scientists, who dropped whatever they were doing and started to look for new evidence bearing on the extinction event." Not only geologists, but paleontologists, chemists, physicists, meteorologists, and even statisticians, were drawn into different aspects of the problem. In the next ten years, more than two thousand scientific papers on the impact hypothesis were presented and published.[9]

Alvarez's own work centered on the search for an impact crater. Finally, in 1991, his decade-long hunt ended. The crater, concealed by millennia of accumulated sediment, lay on the Yucatán coast; and it was *125 miles across.*

A crater that size implied a striking object about 10 kilometers in diameter—as big across as San Francisco, taller than Mount Everest. Its impact would have vaporized crust, set forests on fire, sent tsunamis ripping through the oceans, and thrown enough debris into the atmosphere to block the sun's rays and create storms of poisonous acid rain. The impact, Alvarez concluded, changed the face of the planet—and wiped out the dinosaurs.[10]

He did not convert the entire scientific world. A healthy subsection of respected paleontologists, led by Alvarez's Berkeley colleague William Clements, argued (and are still arguing) for the slow, complex decline of the dinosaur population, not wholesale extinction stemming from a single event. But, like continental drift, the impact hypothesis provided a simple, elegant explanation for a whole range of strange phenomena, stretching across several different scientific fields.

And science isn't immune to a good story. Lyell's long, steady history was not a particularly gripping one, and reintroducing catastrophism brought a bit of flair (not to mention melodrama) back to the field. In 1997, Alvarez published his account of the hypothesis's formation in *T. rex and the Crater of Doom.* For the most part a carefully written, precise account of the clues that led Alvarez and his team to their conclusions, the book begins with a first chapter called "Armageddon," a quote from *The Lord of the Rings*, and a dramatic account of what the impact must have looked like ("Entire forests were ignited, and continent-sized wildfires swept across the lands. . . . Even as the forests were set ablaze, another horror was approaching the coasts.") As science writer Carl Zimmer puts it, "Suddenly the history of life was more cinematic than any science fiction movie."[11]

This cinematic history didn't just belong to geology; catastrophism opened a whole range of scientific fields back up to the possibilities of extraordinary events. Comets, asteroids, supernovas, abnormal flares from the sun, supervolcanoes—each first entered the scientific world, and then popular culture (and then ended up as the Syfy movie of the week).

"A series of catastrophes has brought each of us to our present state," the Harvard astrophysicist Robert Kirshner was able to write in 2002. "The calcium and iron atoms that form our bones and blood were forged in the crucibles of stellar catastrophes." It was now possible to introduce the onetime catastrophe into the history not just of the earth, but of the cosmos itself.[12]

WALTER ALVAREZ

T. rex and the Crater of Doom

(1997)

Alvarez's readable and enlightening account is available in both print and digital formats.

Walter Alvarez, *T. rex and the Crater of Doom*, Princeton University Press (paperback and e-book, 2008, ISBN 978-0691131030).

PART

IV

READING LIFE

(With Special Reference to Us)

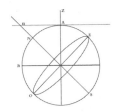

Jean-Baptiste Lamarck, *Zoological Philosophy* (1809)

Charles Darwin, *On the Origin of Species* (1859)

Gregor Mendel, *Experiments in Plant Hybridisation* (1865)

Julian Huxley, *Evolution: The Modern Synthesis* (1942)

James D. Watson, *The Double Helix* (1968)

Richard Dawkins, *The Selfish Gene* (1976)

E. O. Wilson, *On Human Nature* (1978)

Stephen Jay Gould, *The Mismeasure of Man* (1981)

NINETEEN

Biology

The first systematic attempt to describe the history of life

Life and organisation are . . . purely natural phenomena,
and their destruction in any individual is also a natural
phenomenon, necessarily following from the first.
—Jean-Baptiste Lamarck, *Zoological Philosophy*, 1809

Summer 1761. The Comte de Buffon was hard at work on the later
volumes of his massive encyclopedia; James Hutton was tramping
through the Scottish highlands, examining salt mines and coal pits;
Bishop Ussher's six-thousand-year-old history of the earth still, for
the most part, ruled.

And France and England were, yet again, at war.

The ancient hostility between the two countries had been gal-
vanized by competition for North American colonies. Across the
Atlantic, British colonists, including a young George Washington, were carrying on a messy forest fight against the French and
their Native American allies; in Europe, Great Britain and its last
remaining major ally, the German kingdom of Prussia, were bat-
tling against the united front of France, Spain, Austria, and Russia.[*]

Seventeen-year-old Jean-Baptiste de Monet was small for his

*The two-continent war had at least two names; the European fight is generally
known as the Seven Years' War, while the North American conflict is more usually called the French and Indian War or the War of the Conquest.

age, but filled with French pride and a very young man's sense of indestructibility. His father had died the year before, leaving him under the indifferent supervision of his ten older siblings; the French army, currently fighting against the Prussians in the German duchy of Westphalia, needed help; and so Jean-Baptiste found an ancient horse and rode off, ready to fight. He arrived at the front just in time for a major attack against the nearby Prussian-English camp.

On the evening of July 15, his division stormed through the great Teutoburg Forest, fell on the nearest wing of the enemy, and was slaughtered. By the following day, over five thousand French soldiers were dead or had been taken prisoner; Jean-Baptiste and thirteen of his companions were the only survivors of his entire regiment.[1]

The shaken teenager was rewarded with a commission as a lieutenant, but the following year a playful wrestling match with a companion apparently dislocated his neck, leaving him once again on the edge of death. A complicated operation, followed by months of convalescence, saved his life but left him both frail and poverty-stricken. He went to work for a banker and in his spare time pursued new interests: medicine first, then the study of living things, and finally, the nature of life itself.

His first research project, a patriotic effort to identify all of the plants of France, drew the attention of the elderly Comte de Buffon, who nominated him for a post at the royal gardens. Over the next few years, as he continued his work in botany, Jean-Baptiste de Monet drew closer and closer to the core of his real interests: the definition of life, the inevitability of death, and the intertwined relationship of the two.

He also began to sign himself as the Chevalier de Lamarck, a title that would normally have belonged to his oldest brother. There is no clear timeline of how this happened, or why. (Perhaps, one nineteenth-century commentator theorized, all ten older siblings had died?) But Lamarck, like his theory, was developing and changing over time. His reinvention was successful; he is now known, universally, as Lamarck.

Hints of Lamarck's developing thought appeared in his first

nonbotanical publication, 1801's *Système des animaux sans vertèbres*. This work dealt with the animals that Aristotle had described as "nonsanguinous," or bloodless. Lamarck observed, accurately, that bloodless animals had no vertebrae, and so invented the modern term *invertebrate*. But the most groundbreaking observation in the *Système* came in the closing appendix, called "On Fossils." Fossilized remains, Lamarck wrote, were signposts to "the state of the revolutions that different points on the surface of the globe have undergone . . . [and] of the changes that living beings have themselves successively experienced."[2]

Post-Buffon, it was hardly revolutionary to suggest that the earth had undergone changes. But changes in *living* creatures—that was a different story. Up until this point, most natural historians had treated animals and plants as coming late to the surface of the globe, arriving more or less already in their present forms. Even the most cutting-edge work on living creatures, Carolus Linnaeus's 1735 *Systema naturae*, dealt only with the existing characteristics of creatures—not their changes over time. But now Lamarck was marrying the history of life to the history of the globe. As the planet altered, so did the creatures on its surface.

In his work *Hydrogéologie*, published the following year, Lamarck elaborated on this idea. Changes in the earth and in the life upon it were intimately related, but they were, nevertheless, separate fields of study. The natural historian, he proposed in his preface, should think of his studies as falling into three areas: the globe itself (the new field of geology); the skies; and living entities. This last field he gave a new name—*biologie*.[3]

Beginning with Aristotle, various attempts had been made to categorize living things; animals, by their physical qualities, their habits, their food; plants, by their structure, their appearance. Lamarck had a much more fundamental division in mind: living (the subject of biology) and not living. Everything on earth was either organic or inorganic, alive or dead.

This distinction required a definition of "living," and Lamarck had long been contemplating one. A living body, he mused in his private papers, is naturally occurring, "organized in its parts," and, by its nature, "limited in its duration." Anything that possesses life is "necessarily doomed to lose it, that is, to suffer death." Inorganic materials were immortal; living creatures were, without exception, sentenced to death.[4]

Another half decade of study and writing led to the publication, in 1809, of Lamarck's masterwork: *Philosophie zoologique*, or *Zoological Philosophy*, a natural history of life. "We may include what essentially constitutes life in the following definition," Lamarck writes,

> Life . . . is an order and state of things which permit of organic movements; and these movements . . . result from the action of a stimulating cause which excites them. . . . Living beings . . . possess, as everyone knows, the faculties of alimentation, development, [and] reproduction, and they are subject to death.[5]

In other words, all living things have movement that originates from within. They also respond to outside stimuli by changing and shifting. And, perhaps most of all, they die.

So the study of *biology* (the *logos*, word, about *bios*, all living things) did not have the luxury of ignoring beginnings and ends—unlike geology. Movement is change; birth is change; death is change. The biologist could not simply describe. He had to explain the presence, and purpose, of change.

Lamarck proposed three principles of change.

The first, the "principle of use and disuse," incorporated decay and death into the forward progress of life itself. Returning to the idea that the history of the earth and the history of life are intertwined, Lamarck theorized that living creatures transform as they respond to shifts in the earth itself:

> The continued use of any organ leads to its development, strengthens it and even enlarges it, while permanent disuse of any organ is injurious to its development, causes it to deteriorate and ultimately disappear if the disuse continues for a long period through successive generations. Hence we may infer that when some change in the environment leads to a change of habit in some race of animals, the organs that are less used die away little by little,

while those which are more used develop better, and acquire a vigour and size proportional to their use.[6]

Small changes in the environment lead to small changes in life; "a great and permanent alteration in the environment . . . induces new habits," which over time produce great changes.

Second, these alterations happen over great periods of time and are brought about by no agency apart from nature itself. Like Hutton, twenty years before, Lamarck rejected the possibility of ancient deluges and comets: "But why are we to assume . . . a universal catastrophe," he writes, "when the better known procedure of nature suffices to account for all the facts which we can observe?"

> With regard to living bodies . . . nature has done everything little by little and successively. . . . In all nature's works nothing is done abruptly, but . . . she acts everywhere slowly and by successive stages. . . . There is no necessity whatever to imagine that a universal catastrophe came to overthrow everything, and destroy a great part of nature's own works.

He acknowledges the part that a "Supreme Author" must have played at the beginning of all things, but insists that this "infinite Power" designed nature to carry out change without further divine interference: "Nature herself," he writes, "has created organisation, life, and even feeling. . . . Nature possesses the necessary powers and faculties for producing herself."[7]

And finally, all of this change has a particular direction—from the simple to the complex, from lesser to greater, from primitive to the most advanced. Life began, long ago, in water and in simplicity, and then evolved forward:

> Water is the true cradle of the entire animal kingdom. . . . It is exclusively in water or very moist places that nature achieved . . . those direct or spontaneous generations which bring into existence the most simple organized animalcules, whence all other animals have sprung in turn. . . . After a long succession of gen-

erations these individuals, originally belonging to one species, become at length transformed into a new species distinct from the first.

This transformation turned simple, "imperfect" living bodies into "the most perfect . . . having the most complex organisation." Loss, death, and decay had a purpose: nature's path led, ultimately, to perfection.[8]

It was a grand theory, but Lamarck was forced to argue for it on the traditional basis of internal consistency. His proposal wasn't exactly open to experimental proofs; the best he could do was to put forward his observations that existing living creatures appeared to be perfectly adapted to their early environments, along with the certainty that the earth had changed enormously over time. The logical conclusion was, obviously, that living creatures had also changed enormously over time.

The great weakness of the proposal was its missing mechanism. *How* did those changes get passed along from generation to generation? Lamarck had no idea (and neither did anyone else), so he was forced to rely on vague Platonic language about nature's "will" to produce transformations. "Something that might amuse the imagination of a poet," sniffed Georges Cuvier, who found the principle of use and disuse entirely ridiculous.[9]

This missing mechanism ultimately doomed the theory; it drew more opposition than agreement, more scorn than acceptance. Lamarck, never in the top rank of French scientists either socially or by training, grew increasingly bitter over the reception of the *Zoological Philosophy*. He continued to defend it, but the frailty that had haunted him since his teenaged injury had progressed; his eyes were failing, he was soon unable to leave his house easily, and his mind increasingly circled around his failures. He had never been wealthy, and his savings shrank to nothing. His wife died. One of his sons had been born deaf and mentally disabled, and he was forced to confine another to an insane asylum. At his death, twenty years after the publication of his groundbreaking theory, Lamarck's two surviving daughters were too poor to bury him properly; his body was deposited into a general pit that was cleared out every five years, with all the bones piled together in an underground catacomb. Today, his final resting place is unknown.

He had been a little too far ahead of his time, a little too grand in his scheme, a little too dismissive of the need for physical evidence. But he provided a pattern for future biologists, an outline that every subsequent natural historian would build on: the first coherent narrative about the history of life. Fifty years later, the prominent biologist Ernst Haeckel—supporter and popularizer of Charles Darwin, and a best-selling author in his own right—finally provided Jean-Baptiste Lamarck with a proper elegy: "To him will always belong the immortal glory," Haeckel concluded, "of having for the first time worked out the theory of descent, as an independent scientific theory of the first order, and as the philosophical foundation of the whole science of biology."[10]

To read relevant excerpts from the Zoological Philosophy, *visit http://susanwisebauer.com/story-of-science.*

JEAN-BAPTISTE LAMARCK
Zoological Philosophy
(1809)

Lamarck's prose style is clear but repetitive; the *Zoological Philosophy* is about five times as long as it needs to be. You can grasp the basics of his argument by reading the Preface, Preliminary Discourse, Chapters 1–4 and Chapter 7 of Part I, and Chapters 1 and 2 of Part II.

The 1914 translation by Hugh Elliot is still standard. It is widely available as an e-book and has also been reprinted (in an exact replication of the e-book) as a paperback by Forgotten Books.

Jean Baptiste Lamarck, *Zoological Philosophy: An Exposition with Regard to the Natural History of Animals,* trans. Hugh Elliot, Macmillan (e-book, 1914; paperback reprint by Forgotten Books, 2012; no ISBN).

TWENTY

Natural Selection

The first naturalistic explanation for the origin of species

Species have changed, and are still slowly changing by the
preservation and accumulation of successive slight favourable
variations.

—Charles Darwin, *On the Origin of Species*, 1859

The *Zoological Philosophy*, sweeping and comprehensive though it
was, left a colossal question unanswered.

From "the most simple organized animalcules," Lamarck had
written, "all other animals have sprung in turn." But how did this
happen? Had all species on earth sprung from a single kind? If so,
what prompted them to divide and differentiate? And for that matter, what *was* a species?

This wasn't a simple question. Lamarck, more interested in life
itself than in its subdivisions, had skirted it with a one-line definition ("Any collection of like individuals which were produced by
others similar to themselves"). But in this, he was not out of line
with his contemporaries. No natural historian, from Aristotle on,
had really offered a satisfactory definition of *species*—or a coherent
explanation of how different species came to be.[1]

Aristotle had grouped animals together using several different
criteria—anatomy, diet, and (most of all) habits. His "species" were
separated by the ways in which each one had adapted to a specific
manner of life. So his divisions were complicated and overlapping:
aquatic and nonaquatic, with feet and without feet, stationary and
free, bloody and bloodless.[2]

Medieval efforts to classify living things tended to follow this pattern by grouping plants and animals together by their structure, or appearance, or habits, or all three simultaneously. "Plants are divided into three groups," writes Abu al-Dinawari, in the ninth-century *Book of Plants*:

> In one, root and stem survive the winter; in the second the winter kills the stem, but the root survives and the plant develops anew from this surviving rootstock; in the third group both root and stem are killed by the winter, and the new plant develops from seeds scattered in the earth. All plants may also be arranged in three other groups; some rise without help in one stem, others rise also but need the help of some object to climb, whilst the plants of the third group do not rise above the soil, but creep along its surface and spread upon it.[3]

Seven hundred years later, the English naturalist Thomas Moufet was still making use of the same imprecise methods to "classify" insects. "Some are green, some black, some blue," he wrote, in the *Theatrum insectorum*. "Some fly with one pair of wings, others with more; those that have no wings they leap, those that cannot either fly or leap, they walk; some have longer shanks, some shorter. Some there are that sing, others are silent."[4]

But as the sixteenth century gave way to the seventeenth, explorers ranged farther and farther into unfamiliar lands; colonists farmed strange grounds and hunted exotic animals through unknown forests; and the relatively small number of known animals, plants, and insects exploded into a panoply of varieties. A better method of organization was needed—along with a more reliable system of dividing all of that organic abundance into groups.

In 1735, Carolus Linnaeus's *Systema naturae* made the first truly scientific attempt to classify living things. After this first version of *Systema naturae* came out, Linnaeus, a Swedish physician and botanist, revised and reworked it for thirty years, publishing the twelfth revised edition not long before his death. He followed medieval custom by grouping plants into one kingdom, animals into another, and minerals into a third (a division still preserved for us in the game Twenty Questions). But his taxonomy departed from medieval practice in its precision and rigor: every living creature was assigned, on the basis of its morphology (form or shape), into a single genus, a single order, and a single class.

Despite its groundbreaking precision, the *Systema naturae* shared with Aristotle, and with every other natural historian who had tried to put living things into categories, the most basic of assumptions: Species were *different*.

For Aristotle, they were *essentially* different, separate in their very being. Medieval thinkers agreed. A species, according to Augustine of Hippo, was *similia atque ad unam originem pertinentia*, "similar and of a single origin," each one created individually and apart from the other. Linnaeus's species are just as static: "We count as many species," he writes, "as different forms were created in the beginning." Even Lamarck, for all his talk of simple animalcules developing into more complex ones, seems to have had in mind a whole *array* of simple animalcules, each brought to life by spontaneous generation and then developing into its own more complex forms.[5]

Species did not develop from each other. They were, as Ernst Mayr put it, fixed, permanent, and bridgeless. And the precision of Linnaeus's classifications only cemented this belief more firmly into place.

·

Three decades after Linnaeus's death, Charles Darwin was born in Shropshire, England, the fifth child of a prosperous physician. It was 1809; the intermittently mad George III ruled Britain, in spells; Jean-Baptiste Lamarck had just published the *Zoological Philosophy*, James Hutton's uniformity was slowly enfolding Bishop Ussher's young earth, and Georges Cuvier was hard at work drafting his alternative theory of catastrophes.

"I was a born naturalist," Darwin later remarked; his childhood was devoted to collecting, fishing, tracking, and reading natural history. But his father sent him first to the University of Edinburgh to study medicine, and then to Cambridge, hoping to launch him into the church. Neither field interested him ("my time was wasted," he wrote, "I was . . . sickened with lectures") and he did more riding than studying, more bird-watching than Greek. "No pursuit at Cambridge," he recollected, "gave me so much pleasure as collecting beetles."[6]

Darwin finished school in 1831 with a decent degree, an encyclopedic knowledge of the natural world, and absolutely no interest in either healing or preaching. But he had impressed several of his Cambridge professors with his extracurricular studies. One of them, the botanist John Henslow, recommended him to another acquaintance, naval officer Robert Fitzroy, as the perfect addition to Fitzroy's upcoming expedition—a two-year sea voyage that would take a complete geographic survey of the South American coast.

Darwin accepted at once. The planned Christmas departure of Fitzroy's ship, the HMS *Beagle*, was delayed when the entire crew got sloshed: "A beautiful day," Darwin recorded in his diary on December 26, "& an excellent one for sailing—the opportunity has been lost owing to the drunkedness & absence of nearly the whole crew." Finally, the *Beagle* set off from Plymouth Sound on December 27, 1831.[7]

The two-year journey extended to five, and the *Beagle* continued from the South American coast to the Galápagos Islands, then to Tahiti and Australia, circling the globe before returning home. Darwin kept copious notes on his observations. Again and again, these notes describe his struggle with the problem of *species*.

To start with, the whole concept of a species was still poorly defined. "No one definition has satisfied all naturalists," Darwin wrote, a quarter of a century later, "yet every naturalist knows vaguely what he means when he speaks of a species." And the fixity and permanence of species (whatever they were) required multiple acts of divine creation. So why were European ground beetles, Alpine cave beetles, and American cave fish all sightless? Had each of these species been created, separately, without sight? Turnips, rutabagas, and various gourds all had enlarged stems; should this be chalked up to "three separated yet closely related acts of creation"? Or perhaps these were not separate species, just varieties? But in that case, the present definitions of *species* were all drastically inadequate.[8]

Darwin's questions were only deepened by the vast variations of living creatures that he now saw. Each island of the Galápagos had its own mockingbird; they did not interbreed, and they differed in vital ways, so each might be considered a different species; yet they were also, essentially, alike. How should they be classified? What accounted for their differences, and (even more) their similarities?

"When I was on board the *Beagle*," Charles Darwin later wrote, "I believed in the permanence of species, but, as far as I can remember, vague doubts occasionally flitted across my mind. On my return home in the autumn of 1836 I immediately began to prepare my journal for publication" (the account would be published in 1839 as *Journal and Remarks*, although it is usually now known as *The Voyage of the Beagle*) "and then saw how many facts indicated the common descent of species, so that in July, 1837, I opened a note-book to record any facts which might bear on the question; but I did not become convinced that species were mutable until, I think, two or three years had elapsed."[9]

That notebook was only the first of a series; and all of them were filled with *problems*. In the notebook that Darwin created between July 1837 and February 1838, he wrote, in part,

> Species are constant over whole country?
> Every animal has tendency to change.—This difficult to prove. . . .
> No answer because time short & no great change has happened.
> Unknown causes of change. . . .
> Each species changes. Does it progress?
> Changes not result of will of animal, but law of adaptation.
> There is nothing stranger in death of species than individuals.
> Difficult for man to be unprejudiced about self.[10]

While he was struggling with the species problem, Darwin was also reading the works of fellow natural philosophers: borrowing some of their principles, rejecting others. Charles Lyell had published the *Principles of Geology* just as Darwin was setting out on the HMS *Beagle*; "I had brought with me the first volume . . . which I studied attentively," Darwin notes, "and the book was of the highest service to me in many ways." He found Lyell's long-and-slow philosophy of change entirely convincing and adopted it for his own ("Natura non facit saltum," he wrote—*Nature does not make sudden jumps*) but disagreed with Lyell's insistence that changes have no particular progression forward. Darwin read Lamarck's *Zoological Philosophy* and appreciated Lamarck's vision of adaptation leading toward more complex, more "perfect" forms—although he made vigorous marginal notes criticizing the theory of use and disuse. "It is absurd this way," he scribbled, "he assumes the want of habit causes animals annihilation of organ and vice versa."[11]

In the fall of 1838, he picked up the most recent edition of Thomas Malthus's best-selling *Essay on the Principle of Population*. Malthus, a professor of history and political economy at the East India Company's training college for its administrators, had first published the essay in 1798, and had been refining it ever since. The future of the human race, Malthus argued, was shaped by two factors:

First, That food is necessary to the existence of man.

Secondly, That the passion between the sexes is necessary and will remain nearly in its present state.

In other words, humanity has an innate drive to reproduce, which means that the population constantly increases. But because the food supply does not increase as rapidly as the population, a large percentage of those born will always die of starvation: the "difficulty of subsistence" provides a "strong and constantly operating check on population."[12]

This, for Malthus, meant that there would never be such a thing as a perfect society, in which all members live "in ease, happiness, and comparative leisure"; some part of the human race will *always* be suffering from poverty and hunger. But Darwin was at once gripped by another thought. "It at once struck me," he later wrote, "that under these circumstances favourable variations would tend to be preserved and unfavourable ones to be destroyed. The result of this would be the formation of new species."[13]

Darwin had found, he believed, the key to the species problem. He mulled it over for some time and in June of 1842 began to work on setting it down in writing. By 1844 he had completed a first draft of the essay that would become *On the Origin of Species*; a few years later he added to this draft the idea that these variations come about as living creatures adapt to the "economy of nature"—the environment around them.

But he was not yet ready to publish his argument; and he was still perfecting it when, in 1858, he received a letter from the British explorer Alfred Russel Wallace. Wallace, fourteen years younger than Darwin, had greatly admired Darwin's *Journal and Remarks*. He had followed Darwin's example and taken a field trip abroad—in his case, first to the Amazon rain forest, and then to the East Indies. He had collected his own observations on tens of thousands of different species and had come to a novel conclusion: species change, become different, *evolve*, because of environmental pressures.

Wallace, then in Indonesia, had been forced by a recurrent fever to spend hours every day lying down. "I had nothing to do but think," he later wrote, and one day

> something brought to my recollection Malthus's "Principles of Population," which I had read about twelve years before. I thought of his clear exposition of "the positive checks to increase"—disease, accidents, war, and famine—which keep down the population of savage races to so much lower an average than that of more civilized peoples. It then occurred to me that these causes or their equivalents are continually acting in the case of animals also. . . . Vaguely thinking over the enormous and constant destruction which this implied, it occurred to me to ask the question, Why do some die and some live? And the answer was clearly, that on the whole the best fitted live. From the effects of disease the most healthy escaped; from enemies, the strongest, the swiftest, or the most cunning; from famine, the best hunters or those with the best digestion; and so on. Then it suddenly flashed upon me that this self-acting process would necessarily improve the race, because in every generation the inferior would inevitably be

killed off and the superior would remain—that is, the fittest would survive.[14]

Wallace jotted these insights down into a quick essay, "On the Tendency of Varieties to Depart Indefinitely from the Original Type," and enclosed it in his letter to Darwin, asking him to pass it on to Charles Lyell, or any other natural philosophers who might find it interesting.

Darwin was gobsmacked: "This essay," he exclaimed, "contained exactly the same theory as mine." As requested, he sent it along to Lyell ("I never saw a more striking coincidence," he wrote, in his cover letter, ". . . all my originality, whatever it may amount to, will be smashed"), along with a short abstract of his own work in progress.[15]

Lyell and his colleague Joseph Hooker, director of the Royal Botanical Gardens and a personal friend of Darwin's, presented both works to the members of the Linnean Society of London, a century-old club for the discussion of natural history; in August of 1858, Wallace's and Darwin's theories were published side by side in the Linnean Society's printed proceedings.

This was the first articulation of the theory of evolution by natural selection. It was a watershed moment in natural history, but apparently no one noticed. The president of the Linnean Society famously remarked, in his annual report for 1858, "The year . . . has not, indeed, been marked by any of those striking discoveries which at once revolutionize . . . science."[16]

The following year, Darwin, energized by Wallace's codiscovery of the principle of natural selection, finally published his entire argument. This first edition—*On the Origin of Species by Means of Natural Selection, or the Preservation of Favoured Races in the Struggle for Life*, laid out a series of arguments, all supporting Darwin's main conclusion: that life, no less than the earth itself, is changing constantly, and that natural causes *alone* account for that change. He had solved the species problem to his own satisfaction: Species were *not* permanent, fixed, *or* bridgeless. They appeared when *previous* species developed variations and those variations proved helpful in the fight for survival.

On the Origin of Species immediately sold out. The book was widely discussed, widely criticized, widely praised and condemned: "The reviews were very numerous," Darwin remarked later; "for a time I collected all that appeared . . . but after a time I gave up the attempt in despair." In 1864 the well-known biologist and philosopher Herbert Spencer used the phrase "survival of the fittest" to describe Darwin's theory; the phrase soon became inextricably entwined with Darwin's work.

Over the next two decades Darwin revised the *Origin of Species* five times. Even in his final revision, he did not take the theory to its logical end; but he had already privately concluded that his principles of natural selection applied to the human race as well. "As soon as I had become . . . convinced that species were mutable productions," he wrote in his later *Autobiography*, "I could not avoid the belief that man must come under the same law."[17] In 1871 he finally published *The Descent of Man*, an extension of his evolutionary principles to the human race.

The *Descent* brought the full implications of the *Origin of Species* into plain sight.

Charles Darwin had put biology on a collision course with the human race's most cherished idea about itself: its uniqueness. "The question raised by Mr. Darwin as to the origin of the species," one reviewer wrote, "marks the precise point at which the theological and scientific modes of thought come into contact. . . . We are brought face to face in this book with the difficult problems which previously had only revealed themselves more or less indistinctly on the dim horizon."[18]

Those difficult problems were now in plain sight—and would remain there.

CHARLES DARWIN
On the Origin of Species
(1859)

On the Origin of Species is widely available in many different editions and formats. Check the textual notes; the original 1859 text is the clearest, most succinct, and most easily grasped by the general reader. Many editions of the 1859 text also reproduce the essay that Darwin added to the third (1861) edition, "Historical Sketch of the Progress of Opinion on the Origin of Species," which lays out his intellectual debt to Lyell, Lamarck, and others; it is brief and worth your time.

The following recommended edition is simply one of many available, but it does reproduce both the 1859 text and the historical sketch.

Charles Darwin, *The Origin of Species*, Wordsworth Editions (paperback and e-book, 1998, ISBN 978-1853267802).

TWENTY-ONE

Inheritance

The laws, and mechanisms, of heredity revealed

By this process the species *A* would change into the species *B*.
—Gregor Mendel, *Experiments in Plant Hybridisation*, 1865

Charles Darwin was sure that variations were passed from parent to child, but he had no idea how this worked. "The laws governing inheritance are quite unknown," he lamented in the second chapter of the *Origin of Species*. "No one can say why a peculiarity . . . is sometimes inherited and sometimes not so."[1]

The most widely accepted nineteenth-century model of inheritance, "blending," posed enormous problems for natural selection. It proposed that the characteristics of both parents somehow passed into their offspring and melded together to create a happy medium: a black stallion and a white mare should have a grey foal, a 6-foot father and 5-foot mother should produce a child who would mature at 5 feet 6 inches. There were two problems with blending, though: First, it was (often) demonstrably untrue. Second, it tended to remove variation, not preserve it.

Nine years after the *Origin of Species* was first published, Darwin suggested that inheritance could be explained instead through the existence of "minute particles" called *gemmules*, which were thrown off by every part of an organism, accumulated in the sex organs, and were then passed on to offspring. The strongest argument for this theory was simply that he couldn't think of anything better. "It is a very rash and crude hypothesis," he wrote to his friend T. H. Huxley, "yet it has been a considerable relief to my mind, and I can hang on it a good many groups of facts."[2]

He never came up with a better explanation, although the key to the truth was literally under his own roof.

At Darwin's death in 1882, his library contained unopened copies of a short paper in German by the Austrian botanist (and Augustinian friar) Gregor Mendel, which Mendel had presented in 1865 to his local natural-history society. The paper recounted Mendel's nine years of experimenting with hybridization; he had tried to create a new species by interbreeding thirty-four different varieties of sweet peas, and although he had failed, he had discovered a series of laws that seemed to govern how the characteristics of the sweet peas (shape and color of seeds and pods, position of flowers, length of stem) were passed on.[3]

These laws, Mendel noted, operated with "striking regularity." Some of the characteristics of the peas were *always* passed on to the next generation, "entire, or almost unchanged"; he called these "dominant" characteristics. Other aspects seemed to disappear in the offspring but then would sometimes reappear unchanged several generations on; these, which became latent, Mendel termed "recessive."

The painstaking cross-fertilization of generation after generation of sweet peas allowed Mendel to work out a series of formulas for the passing on of these dominant and recessive characteristics. Clearly, the characteristics were carried from parent pea to offspring pea by the egg and pollen cells, so (Mendel proposed) those cells must contain discrete units, or *elements*, with each element carrying a particular characteristic within it: "The differentiating characters of two plants," he concluded, "can finally . . . only depend upon differences in the composition and grouping of the elements which exist in the foundation-cells." The proper manipulation of those elements could change the characteristics of the next generation—and, eventually, even mutate one species into another.

If a species *A* is to be transformed into a species *B*, both must be united by fertilisation and the resulting hybrids then be fertilised with the pollen of B; then, out of the various offspring resulting, that form would be selected which stood in nearest relation to B and once more be fertilised with B pollen, and so continuously until finally a form is arrived at which is like B and constant in its progeny. By this process the species A would change into the species B.[4]

This was the answer to Darwin's vexing problem, the mechanism by which a variation could pass from one generation to the next and ultimately shape a species into something else.

But a cascading series of events prevented Mendel from developing his research further. He was, first and foremost, a friar rather than a scientist, so he did not push for the translation and distribution of his paper; it was known, mostly, to the forty people who had bothered to turn up in 1865 to hear him read it. The one prominent scientist who did take notice of it, the Swiss botanist Karl Wilhelm von Nägeli, was a die-hard proponent of blended inheritance and sharply criticized Mendel's results. Nägeli insisted that Mendel redo all of the experiments using hawkweed, but this venture failed horribly because hawkweed does not need to be pollinated in order to produce seeds (meaning that none of the resulting seeds were actually hybrids). And in 1868, Mendel's monastery elected him abbot for life; this not only pulled him away from his experiments but got him embroiled in a complicated, time-consuming argument with the Austrian government over the amount of tax the monastery paid.

Still, Mendel continued to think of himself as a naturalist; he went on experimenting with breeding and hybridization of flowers, fruit trees, and honeybees, and he became known, over the next decade, for his meteorological observations and theories. "My scientific work has brought me a great deal of satisfaction," he told a colleague in 1883, the year before his death, "and I am convinced that it will be appreciated before long by the whole world."[5]

He probably wasn't thinking about the sweet-pea research, which had been eclipsed by his other concerns; but the sweet peas were what the world remembered.

•

Just one year after Mendel's paper was published, the German biologist Ernst Haeckel proposed that inheritance might be controlled by something deep inside the core of the cell. He didn't
have either the equipment or data to prove this theory, but two
decades later, his countryman Walther Flemming made use of
much-improved microscopic lenses and better staining techniques
to observe minuscule, threadlike structures in cells that had begun
to divide (mitosis). Flemming's colleague Wilhelm Waldeyer suggested that these should be called *chromosomes*, a name that simply
described their ability to soak up dye (*chroma*, "color"; *soma*,
"body").[6]

And then the sweet peas resurfaced.

On May 8, 1900, the botanist William Bateson, was on his way
to deliver a talk about the difficulties of heredity to the Royal Horticultural Society of England. He had brought a sheaf of research
articles to read on the train; Mendel's German paper was among
them. By coincidence, Bateson's hand lighted on Mendel's work
first; halfway through, he realized that the formulas were the key
to the problem of inheritance. (Possibly he had read the paper earlier, but the train story is the one he always told afterward; it had
more dramatic flair.)

Coincidentally, two other researchers had also recently come
across Mendel's work: the Dutch botanist Hugo de Vries, who had
been conducting his own experiments with hybridization; and a
student of Nägeli's, Carl Correns. Within the year, all three men
published papers dealing with problems of heredity and citing
Mendel's "laws" of inheritance. In 1901 the Royal Horticultural
Society sponsored the first translation of the German study into
English, bringing Mendel's laws fully into the public gaze.[7]

All that remained was to connect the chromosomes with the
laws.

In 1902 the German biologist Theodor Boveri carried out a
series of experiments demonstrating that sea urchin embryos
needed exactly thirty-six chromosomes to develop normally—

which strongly suggested that each chromosome carried a unique and necessary piece of information. Simultaneously, American graduate student Walter Sutton, working at Columbia University, concluded from his experiments with grasshoppers that chromosomes carry the "physical basis of a certain definite set of qualities." Not Darwin's gemmules, or Mendel's elements: *genes*, carriers of information from one generation to the next.[8]

The term was coinvented by William Bateson, who first called the new study of chromosomes and their relationship to inheritance *genetics*, and the Danish botanist Wilhelm Johannsen, who separated out the word *gene* and applied it to the unit of heredity itself. Forty years after Mendel first published his experiments, twenty years after his death, he created a whole new field within biology.

As he had predicted, the whole world finally appreciated his work.

GREGOR MENDEL
Experiments in Plant Hybridisation
(1865)

Mendel's paper was translated into English by the Royal Horticultural Society of London in 1901; this clear and succinct translation remains the standard. W. P. Bateson's republication of the entire English-language paper in his 1909 book *Mendel's Principles of Heredity* is widely available online; Cosimo has also republished it in a high-quality paperback with all formulas and diagrams included.

Gregor Mendel, *Experiments in Plant Hybridisation*, Cosimo Publications (e-book and paperback, 2008, ISBN 978-1605202570).

TWENTY-TWO

Synthesis

*Bringing cell-level discoveries
and the grand story of evolution together*

Biology at the present time has embarked upon a period of
synthesis after a period in which new disciplines were taken
up in turn and worked out in comparative isolation. . . .
Already we are seeing the first-fruits in the re-animation of
Darwinism.
— Julian Huxley, *Evolution: The Modern Synthesis*, 1942

World War I was over, and Charles Darwin was in trouble.

These were not unrelated phenomena. "The one event . . .
which more than any other single one laid the foundation for the
situation in which Western Europe finds itself today," the American biologist Raymond Pearl remarked, just before the armistice of
1918, "was the publication of a book called *The Origin of Species*."
He was voicing a widespread opinion: Darwinian natural selection, by removing the human race from its special position as a
purposeful and unique creation of God, had plunged humanity
into the same grim, amoral struggle for survival that the rest of the
animal kingdom endured. "All nature is at war," Darwin himself
had written. It was no wonder that this war had engulfed *Homo
sapiens* too; no wonder, Pearl wrote (in an eerily prescient phrase),
that the Germans had ruthlessly undertaken the "biological elimi-
nation" of another nation.[1]

Theological objections to *The Origin of Species* had also gained steam. "There is no country in the whole world in which the Christian religion retains a greater influence over the souls of men than in America," Alexis de Tocqueville had famously observed, a century before; and in America, the challenge to Darwin took the form of dogged loyalty to the literal truth of the Genesis account. The Butler Act, signed into Tennessee law in 1925, banned the teaching of natural selection in place of special creation, and the Scopes trial upheld the act's legality.[2]

But natural selection was also coming under heavy fire from biologists who were less concerned with its implications for morality and more puzzled by its ramifications for evolution.

Natural selection, working through chance variations, did nothing more than weed out the weak; this seemed to be a completely inadequate explanation for the progression, the increased complexity, the *direction* that seemed part of the history of life. Natural selection seemed far more likely to produce haphazard and unguided change. And so scores of prominent biologists (the Russian Lev Berg, the Austrian Ludwig von Bertalanffy, the German Otto Schindewolf, and many, many others) suggested various kinds of *orthogenesis*, the assurance that *some* predetermined pattern, or goal, or intention, lay behind the origin of species.

Better instruments, more data, and improved research techniques were yielding discoveries thick and fast, many of them (in cytology, biometry, embryology, genetics) suggesting that natural selection was indeed an adequate explanation for organic life. But these studies were clogged with technical language, inaccessible to the nonspecialist. There was, in Ernst Mayr's words, "an extraordinary communication gap between the various disciplines of biology," and an even greater gap between biology and the other sciences. Genetics, in particular, was yielding insight after insight, but geneticists were taking no time to build bridges from their discoveries to the phenomena that fieldwork in natural history had revealed: animal behavior, fossilized remains, the complex interactions of living environments.[3]

"Darwin's selection theory . . . is far from explaining the ultimate causes of numerous adaptations," noted *Lehrbuch der Zoologie*, a standard university text in Germany in the 1920s. Prominent botanist Wilhelm Johannsen was even more emphatic: "[It is] completely evident that genetics has deprived the Darwinian theory of selection entirely of its foundation," he wrote. "We have never been less sure about the mechanisms of evolution," admitted the French biologist Jean Rostand. And in 1937, Paul Lemoine, director of the National Museum of Natural History in Paris, went even further: "Natural selection plays no role," he concluded. "The data of genetics furnish no argument in favor of evolution. . . . The theory of evolution will very soon be abandoned."[4]

.

Julian Huxley seems to have been genetically predestined to rescue Darwin.

His grandfather, Thomas H. Huxley, had been one of the earliest reviewers of the *Origin of Species* and afterward became one of Darwin's most ardent supporters and friends. "*The Origin of Species*," he had written to a friend, a dozen years after the book first appeared, "has worked as complete a revolution in biological science as the *Principia* did in astronomy." All his life, T. H. Huxley remained such a fierce defender of Darwinian theory against its critics that he once nicknamed himself "Darwin's bulldog."[5]

Young Julian, born in 1887, was a naturalist from childhood: interested in "plants and animals, fossils and geography," a frog collector, butterfly classifier, and bird-watcher. "Julian evidently inclines to biology," his distinguished grandfather remarked when the boy was only four years old. "How I should like to train him!" He studied zoology at Oxford and, after his graduation in 1909, remained at the university to conduct experiments in a startling range of topics: embryology, ontogeny, cellular differentiation, morphogenesis, genetics, and (he always combined his laboratory work with field observations) the courtship rituals of redshanks and crested grebes.[6]

Julian Huxley remained in the academic world until the late 1920s: an effective and much-respected teacher and researcher, but restless, increasingly weary of undergraduates, unhappily married, and struggling with bouts of depression and mania. In 1926 the writer H. G. Wells invited him to collaborate on a massive encyclopedia of biology, intended to sum up the entire progress of the field. Wells, twenty years Julian's senior, had studied biology in London under T. H. Huxley himself; he was already famous as the author of *The Time Machine* and *The War of the Worlds* (and many other books).

Julian seized at the opportunity, left his teaching post, and began work. He was a polished writer, but Wells (notoriously demanding and difficult) insisted that he refine his style even further, so that the increasingly complicated developments in biology would be accessible to lay readers. "I had learnt a great deal . . . under H. G.'s stern guidance," Huxley later remarked, "about the popularization of difficult ideas and recondite facts . . . [and about] synthesizing a multitude of facts into a manageable whole, aware of the trees, yet seeing the pattern of the forest. . . . This, I may add, did not come easily."[7]

The Science of Life was an international best seller, but the ability to write well on a very big subject turned out to be even more valuable than the royalties.

An increasing cadre of distinguished scientists—Julian Huxley among them—had begun to recognize the need to draw together new discoveries in genetics and other aspects of natural history into a whole: a Big Story, an explanation of how it all fit together. What was needed was a defense of Darwin that nevertheless took into account the need to bring Darwin's original theory, now well over a half century old, into line with the newest discoveries in genetics and cytology. Both scientists and the public needed a greater understanding of "evolution at work," the ways in which the grand theory and specific discoveries acted together.[8]

So, in the tradition of his grandfather, Darwin's bulldog, Julian Huxley proposed in 1936 that a new society be founded: the Association for the Study of Systematics in Relation to General Biology. He was its first chair; and at its first meeting, on a Friday in June of 1937, seventy-four biologists attended. Among the goals of the new society, reported the journal *Nature*, was to "stimulate discussion and to promote cooperation between workers in different branches of biology."[9]

The Russian entomologist Theodosius Dobzhansky, a regular

correspondent of Huxley's, and a fellow systematist, was already set to publish. *Genetics and the Origin of Species* (1937) brought together Dobzhansky's laboratory work in genetics, his observations in the field (he had worked extensively with fruit flies), and the somewhat obscure (to nonspecialists) mathematical calculations of population genetics—all in order to argue that Darwinian natural selection *did*, indeed, account for the existence of species. It was one of the first systematic, big-picture works in modern biology, but far from the last.[10]

In the next decade, George Gaylord Simpson's *Tempo and Mode in Evolution*, Bernhard Rensch's *Evolution above the Species Level*, and Ernst Mayr's *Systematics and the Origin of Species, from the Viewpoint of a Zoologist* all appeared. And Huxley was hard at work on his own big-picture volume. His *Evolution: The Modern Synthesis* appeared in 1942, and two things set it apart: Huxley was, intentionally, writing for the informed and interested layperson, not simply for his scientific colleagues; and Huxley, for the first time, had used the word *synthesis*.

"The death of Darwinism has been proclaimed not only from the pulpit, but from the biological laboratory," Huxley begins, "but, as in the case of Mark Twain, the reports seem to have been greatly exaggerated, since to-day Darwinism is very much alive." And his first chapter lays out his intentions:

> Biology in the last twenty years, after a period in which new dis-
> ciplines were taken up in turn and worked out in comparative
> isolation, has become a more unified science. It has embarked
> upon a period of synthesis, until to-day it no longer presents the
> spectacle of a number of semi-independent and largely contradic-
> tory sub-sciences, but is coming to rival the unity of older sci-
> ences like physics, in which advance in any one branch leads
> almost at once to advance in all other fields, and theory and
> experiment march hand-in-hand. As one chief result, there has
> been a rebirth of Darwinism. . . . The Darwinism thus reborn is
> a modified Darwinism, since it must operate with facts unknown
> to Darwin; but it is still Darwinism in the sense that it aims at
> giving a naturalistic interpretation of evolution. . . . It is with this

reborn Darwinism, this mutated phoenix risen from the ashes of the pyre . . . that I propose to deal in succeeding chapters.[11]

It was a sprawling, multifaceted task, and *Evolution: The Modern Synthesis* is a sprawling, multifaceted book, covering in turn paleontology, genetics, geographic differentiation, ecology, taxonomy, adaptation, and the idea of evolutionary progress.

But Huxley's training period with H. G. Wells had served him well. The clarity of his style and the down-to-earth, jargon-free presentation of technical ideas made *Evolution* both readable and an instant success. "The outstanding evolutionary treatise of the decade, perhaps of the century," exclaimed the *American Naturalist*, one of the most important journals of the field, and readers agreed. Huxley's book went through five printings and three editions; the latest, in 1973, included a new introduction, coauthored by nine prominent scientists, affirming the overall truth of the synthesis and updating its data.[12]

From 1942 on into the twenty-first century, this entire endeavor—the careful connection of cell-level studies in genetics with the larger world of natural history—would continue, drawing from a wider and wider array of developing subspecialties (such as the late-twentieth-century field of evolutionary genomics). And it would continue to take its name from Huxley's book: the *modern synthesis*. Huxley had resurrected Darwin, and the "mutated phoenix" was evolving steadily forward.

JULIAN HUXLEY
Evolution: The Modern Synthesis
(1942)

In 2010, MIT Press published the 1942 text of *Evolution: The Modern Synthesis* along with Huxley's original preface, as well as the introductions to the second and third editions.

Julian Huxley, *Evolution: The Modern Synthesis: The Definitive Edition*, MIT Press (paperback, 2010, ISBN 978-0262513661).

TWENTY-THREE

The Secret of Life

Biochemistry tackles the mystery of inheritance

Science seldom proceeds in the straightforward logical
manner imagined by outsiders.
—James D. Watson, *The Double Helix*, 1968

For a hundred years—at least since Lamarck—the science of life
had accepted that living creatures pass attributes to their young. But
how this worked was a mystery. *Something* was inherited; but what?
What did it look like; how did it behave; where was it? And how
did this "pangene," this unit of information, go about producing a
similar eye color, or height, or fur pattern, in the next generation?

In 1953 the young American James Watson and his British
colleague Francis Crick, working in Cambridge, discovered the
answer: DNA, the double-helix strands of molecules that replicate
a parent's characteristics in a child. Crick was so thrilled that he
bounded into the nearest pub and announced to all within earshot
that he had just found "the secret of life." This, as scores of science
texts will tell you, was a discovery that "changed the world"; "one
of the most important scientific discoveries of the twentieth century"; "the birthday of modern biology." Fifteen years later, James
Watson guaranteed the immortality of that watershed moment by
writing an instantly popular account of his work on DNA: *The
Double Helix*.[1]

But the existence of deoxyribonucleic acid had been known for
over a century. And its double-helix structure would not actually
be observed for years to come. The "discovery of DNA" was, in

fact, a series of minute advances forward in chemistry, biology, and even physics, rooted in seventeenth-century technology, carried out over the decades that bridged the nineteenth and twentieth centuries, dependent on the work of scores of scientists, and finally channeled by one charismatic researcher into a best-selling popular account.

•

Since Robert Hooke first peered through his microscope at a piece of cork, natural scientists had accepted that living things were made up of separate bits, like a whole honeycomb divided into tiny parcels of sweetness. "The substance of Cork is altogether fill'd with Air," Hooke had written, in *Micrographia*, "and . . . that Air is perfectly enclosed in little Boxes or Cells, distinct from one another." Hooke observed cells in many living things, as far distant as petrified wood and spiders; following his lead, other observers discovered cells in all sorts of vegetables, in embryos, in animal tissues.

What was inside those cells, and why they existed at all (why not simple undifferentiated flesh?), defied eighteenth-century science. But by the late 1830s, making use of improved instruments to extend their senses, two different observers (the French biologist Félix Dujardin, and Jan Purkinje of Bohemia) concluded that Hooke's tiny boxes were filled with a "glutinous, diaphanous substance," sticky and intractable, essential to life: *protoplasm*, as Purkinje called it. Protoplasm was the most basic "material of life," a fundamental jelly whose exact purpose was still unknown.[2]

In 1847, two German naturalists—botanist Matthias Jakob Schleiden and zoologist Theodor Schwann—defined Hooke's *cells* as the first and most fundamental unit of life. In each new living thing, cells grew and developed, starting as tiny grains and expanding outward: "It is an altogether absolute law," Schwann wrote, "that every cell . . . must make its first appearance in the form of a very minute vesicle, and gradually expand to the size which it presents in the fully-developed condition." This cell, confirmed the German biologist Max Schultz in an 1861 paper, was a ball of protoplasm containing a distinct center—a *nucleus* (from the Latin word for "kernel" or "nut").[3]

Meanwhile, chemistry—a field that had been preoccupied with metals, gases, and other inorganic substances ever since Robert Boyle's seventeenth-century elaborations—had been set on an intersecting course with biology. In 1828 the chemist Friedrich Wöhler accidentally produced the organic compound urea, naturally found in urine, in his lab. ("I can make urea without needing a kidney!" he wrote to a colleague in great excitement.) The unexpected synthesis suggested that organic substances could be understood through the same basic chemical laws that were known to govern inorganics. Cells could be prodded, stimulated, catalyzed, broken apart, subjected to chemical tests, understood by means of their chemical reactions: the beginnings of *biochemistry*.[4]

The infant science grew quickly. In 1833, the French chemists Anselme Payen and Jean-François Persoz discovered that *something* in malt (they called it "diastase") had the ability to change starches into sugars. Diastase was the first known *enzyme*—an organic molecule, generally a protein, that sets a chemical reaction into motion and has the potential to change a living thing. Four years later, the Swedish chemist Jöns Berzelius came up with the name *catalysis*, a process "different from [those] previously known to us," which has the power to cause "a rearrangement of the constituents of the body into other relationships." The discovery of catalysis had implications far past the test tube. When, Berzelius wrote,

we turn with this idea to living Nature, an entirely new light dawns for us. It gives us good cause to suppose that in living plants and animals thousands of catalytic processes are taking place between the tissues and the fluids, producing the multitude of dissimilar chemical compounds for whose formation from the common raw material, sap or blood, we had not been able to think of any cause, but which in the future we shall probably find in the catalytic power of the organic tissue.[5]

The ultimate explanations for our biological existence, for our shape and form, lay in *chemistry*—in understanding the reactions that gave rise to our cells and governed their interactions.

But the cell was still an uncharted territory, and neither biologists nor chemists could do much more than guess at its contours.

•

In 1865 the Swiss medical student Friedrich Miescher recovered from typhoid fever, but not without scars: his hearing had been damaged, one ear left completely deaf.

This made patient care difficult, so Miescher decided to occupy himself with medical research instead. He took his lead from an uncle, the distinguished physician Wilhelm His: "I came to the conclusion," His later wrote, in his personal papers, "that the final solution of the problems of tissue development could be solved only by chemistry." Miescher's own chemical interests lay in the exact makeup of the cell—particularly the mysterious nucleus, difficult to glimpse, its function unknown. He had noticed that the nuclei of *lymphoid* (white blood) cells were, observably, larger in proportion than those of other cells, so he decided to collect pus from discarded surgical bandages, isolate the nuclei of the white blood cells (using various solvents, as was the accepted practice), and analyze their makeup.[6]

Miescher's experiments, conducted over a two-year period and published in 1871, revealed an unexpected presence. It was generally believed that nuclei contained proteins, considered to be (in some way) the building blocks of life. But these nuclei broke down into two different parts—a protein, yes, but also a previously unknown, lightly acidic substance. Miescher gave this new acid a name: *nuclein.**

In 1929 the Lithuanian biochemist Phoebus Levene, driven to the United States by anti-Semitism in his own country and now working at the Rockefeller Institute in New York, identified one of the most abundant elements in Miescher's nuclein as containing a sugar called *deoxyribose.* He called this particular acid in the nuclein *deoxyribonucleic acid.*[7]

Miescher had discovered DNA; Levene had named it. Neither man had any idea what it did.

---

*In this context, an "acid" is simply a substance that, when added to water, increases the concentration of protons, or hydrogen ions (H^+), *in* the water.

•

As biochemists dug deeper into the chemical makeup of life, biologists continued to puzzle over Mendel's genes.

A swirl of information surrounded them, but no one had yet snatched the relevant bits from the air and pieced them together. In the early 1880s, Walther Flemming had seen chromosomes, tiny threadlike structures in dividing cells; in 1890 the biologist Hermann Henking had noticed that, while some chromosomes paired off evenly during cell division, like square-dance partners meeting up to sashay down the line, others (apparently identical) made it into only *half* of the newly produced cells. He had no idea why this was so, but he gave the sometimes-tardy chromosome a name: the X chromosome, a mystery.[8]

After the German biologist Theodor Boveri confirmed, in 1902, that chromosomes were somehow responsible for ferrying Mendel's "genes" from generation to generation, the next step was to relate this bit of information to the existence of two different kinds of chromosomes. Three different biologists, all American, added their observations: Clarence McClung, working on grasshoppers, was the first to suggest that perhaps the presence or absence of the X chromosome was responsible for determining the sex of offspring; Edmund Beecher Wilson, studying *hemipterans* (aphids, cicadas, sweet-potato bugs) theorized that the X chromosome had to be present in *female* offspring; and Nettie Stevens, working with higher-quality cells from fruit flies, confirmed Wilson's theory through observation.[9]

After Stevens published her fruit fly results, the zoologist Thomas Hunt Morgan led a seven-year research project at Columbia University that was able to track generation after generation after generation of fruit flies. Fruit flies reproduce so quickly that seven years is practically deep time; with so much information to chart, Morgan and his colleagues were able to show that the genetic information determining eye color was carried on the X chromosome.[10]

This was the first direct association of a particular *phenotype* (an observable quality, such as height, weight, nose shape, hair color) with a specific *genotype* (arrangement of chromosomes). And it provided a biological explanation for something that had been observed for centuries: that some physical qualities cannot be passed directly from father to son.*

The members of one family may bleed profusely, while those of another family may bleed little. So says the Babylonian Talmud, compiled around the third century; it is one of the oldest references we have to hemophilia, the disorder that prevents blood from clotting. In the centuries after, physicians grappled with this strange condition. The tenth-century Cordoban surgeon Albucasis observed that healthy mothers could give birth to hemophiliac sons; at the beginning of the nineteenth century, the Philadelphia doctor John Otto wrote that the disorder seemed to appear only in males; and the royal families of Germany, Spain, and Russia were tormented by the disease's irregular appearance in their princes.[11]

Charles Darwin himself had charted the pattern of inheritance, which looked something like this:

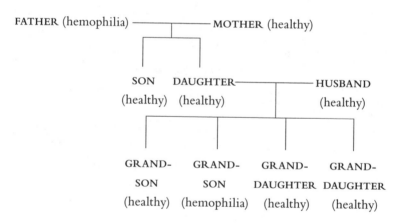

Morgan's fruit fly experiments now made it possible to explain the oddness of the pattern. If the genetic information for hemophilia is carried only on the X chromosome, the disease itself will show up only if *all* copies of the X chromosome are affected. And since male

---

*It should be noted that sex determination works very differently in fruit flies and in mammals, but the fruit fly research provided a theoretical framework for understanding sex-linked characteristics.

children have only one X chromosome, they are far more likely to be afflicted. (Dr. Otto was misled by statistics; it is uncommon, but not impossible, for women to suffer from hemophilia.)

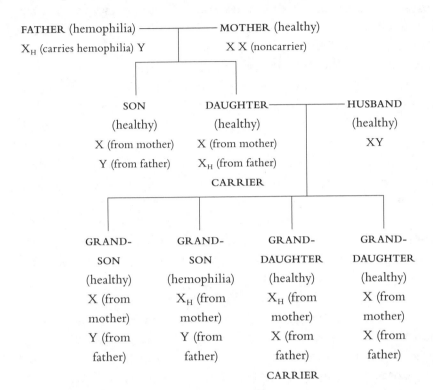

FATHER (hemophilia) ——————— MOTHER (healthy)
X_H (carries hemophilia) Y   X X (noncarrier)

SON — DAUGHTER — HUSBAND
(healthy) — (healthy) — (healthy)
X (from mother) — X (from mother) — XY
Y (from father) — X_H (from father)
CARRIER

GRAND-SON (healthy) — GRAND-SON (hemophilia) — GRAND-DAUGHTER (healthy) — GRAND-DAUGHTER (healthy)
X (from mother) — X_H (from mother) — X_H (from mother) — X (from mother)
Y (from father) — Y (from father) — X (from father) — X (from father)
CARRIER

Morgan's work made it possible to uncover the invisible lines of transmission: an apparently healthy daughter is actually a silent carrier of the disease, with a 50/50 chance of passing it on to an infant son.

But all this still had nothing to do with the *mechanism* of inheritance. How the genetic information was carried on the chromosome, and how it went about shaping and forming the recipient (the relationship between genotype and phenotype, in other words), remained a mystery. Biologists working with chromosomes and genes were still doing little more than keeping statistics: observing which chromosomes went where, and what the result might be.

Light began to dawn when chemistry, biology, and physics formed a brief three-way intersection.

In 1927 the biologist Hermann Muller, a coworker of Morgan's in the fruit fly project, announced that bombarding fruit flies with X-rays *changed* their genetic information. His irradiated flies had produced offspring with a whole array of new phenotypes; Muller himself lists "splotched wing," "white eye," "miniature wing," "forked bristles," and more. The work of Roentgen, the Curies, and Rutherford had already shown that radiation produced *changes* in the structure of atoms and molecules;* Muller's results suggested that the still-mysterious *genes* were, in fact, molecules—structures that were vulnerable to the changes that X-rays produced. And since radiation didn't produce the same changes over and over again, genes were likely to be a whole array of different molecules, rather than a sack of similar particles, "merely repetitions of one another."[12]

So: genes were molecules, conveyed by chromosomes from parent to child. That was an answer to the first part of the puzzle. The second remained: How did a molecule produce a certain shape of earlobe, a long second toe, freckles: *phenotypes*?

The answer came by way of bread mold.

In the 1940s, two Stanford biochemists, named George Beadle and Edward Tatum, carried out a series of experiments on *Neurospora*, a fungus that grows on bread (and also turns soybean scraps and coconut remnants into the Indonesian staple dish known as *oncom*). Their work showed that when genes were altered, the production of certain *enzymes* ceased—making the cells of an organism unable to carry out certain chemical reactions.

Those chemical reactions were what gave the organism its identity. A century before, Anselme Payen and Jean-François Persoz had identified the first enzyme, and Jöns Berzelius had put his finger on the importance of "catalytic processes" for living creatures. Over the decades that followed, biochemists refined and built on this knowledge, so that Beadle and Tatum were now able to define a living thing—its entire makeup, its metabolism and its phenotype, the way it functioned and the way it looked—as the totality of its chemical reactions. Enzymes were the catalysts of

*See Chapter 16.

those chemical reactions. An altered gene resulted in an altered enzyme; an altered enzyme resulted in an altered phenotype; an altered phenotype produced a mold cell that needed different nutrients than its parent, a splotch on a fruit fly's wing, a long second toe or cleft chin. Mess with the enzymes, and you change the organism.

Here at last was the connection between genotype and phenotype. Beadle and Tatum worked for the next few years making maps of which enzymes, carried by which chromosomes, produced different sorts of phenotypes. Their work became known as the "one gene, one enzyme" hypothesis: genes (human, bacterial, or otherwise) affected the production of enzymes, and enzymes affected the characteristics of the living creature.[13]

But the gene itself was *still* a mystery. There was no familiar molecule, no recognized organic substance, no already-identified chemical compound that had such power over enzymes.

Simultaneously, a biologist named Oswald Avery (working, as Levene had, at the Rockefeller Institute in New York) was studying pneumococcal viruses. One particular pneumococcal strain had a unique quality: the viral cells could form capsules around themselves, made out of a complex molecule called a polysaccharide. These capsules strengthened the virus and made it more potent.

Avery realized that if a solution of DNA (the acidic substance identified, by Levene, as an element of Miescher's "nuclein") was taken from a pneumococcal strain capable of forming polysaccharide capsules and introduced into *other* pneumococcal strains, the recipient strains were transformed. Suddenly, they too were able to form polysaccharide capsules.

This could have been a eureka moment.

In fact, Oswald Avery (a cautious, responsible man) mused in a private letter to his brother that this DNA solution appeared to be acting, unexpectedly, like one might expect a "gene" to act. But he and his colleagues were too skeptical of their results to make any great claims. They published their results in the *Journal of Experimental Medicine*, where the paper attracted little interest. The study of cells and their properties had been so spread out across disciplines, fragmented into subfields and minidiscoveries, that few biologists (or biochemists) could claim full knowledge of all the relevant discoveries.[14]

Only a small fraction of those biologists knew anything about a third, parallel but separate investigation: biologists Max Delbrück and Salvador Edward Luria and bacteriologist Alfred Hershey were working together at a Long Island lab, attempting to pull apart the exact structure of viruses that had the ability to infect bacteria. These viruses ("bacteriophages") invaded bacterial cells, reproduced inside them, and then destroyed the host cells and set themselves free.

It became increasingly clear that these viruses had the ability to replicate themselves *exactly*. And in 1947, another biochemist, Seymour Cohen, noticed that when a certain strain of bacteriophage invaded bacterial cells, the synthesis of DNA within the infected cell abruptly stopped—and then began again, a few minutes later, at a multiplied pace. This was a strong hint that DNA had *something* to do with that exact replication.

Other researchers theorized that perhaps DNA was the agent that allowed cells to reproduce themselves. But the proteins in nuclein were generally thought to be a much stronger contender for that role. In neither case could scientists explain exactly how the genetic information was coded by the parent cell, or decoded by the child.[15]

A very young James Watson, still pursuing his PhD and not quite sure what he wanted to study, began to work with the Long Island "phage group" in 1948. At first he had little interest in the study of nucleic acids, a field where "ideas could not be easily disproved": "Much of the talk about the three-dimensional structure of proteins and nucleic acids was hot air," he later observed. Theories abounded, none of them susceptible to tests.

But in 1951, in Europe on a postdoctoral fellowship, Watson attended a talk given by the British physicist and biologist Maurice Wilkins, who had been working on DNA at Kings College in London. Wilkins's talk was illustrated by X-ray pictures of DNA that showed clear structural patterns, and suddenly Watson found himself gripped. Surely, such a clear pattern could be the key to finding the "regular structure" of DNA, and perhaps that structure would demonstrate that DNA was responsible for carrying that elusive genetic information.[16]

Defying the terms of his fellowship, Watson wangled himself a position in England, working at Cambridge's Cavendish Laboratory with the biophysicist Francis Crick. Crick, twelve years his senior, had been interested in DNA for some years, but Watson discovered—to his surprise—that the conventions of English society had prevented Crick from pursuing further study. "At this time," Watson wrote in *The Double Helix*, "molecular work on DNA in England was, for all practical purposes, the personal property of Maurice Wilkins."

> It would have looked very bad if Francis had jumped in on a problem that Maurice had worked over for several years. . . . It would have been much easier if they had been living in different countries. The combination of England's coziness—all the important people, if not related by marriage, seemed to know one another—plus the English sense of fair play would not allow Francis to move in on Maurice's problem. In France, where fair play obviously did not exist, these problems would not have arisen. The States also would not have permitted such a situation to develop. One would not expect someone at Berkeley to ignore a first-rate problem merely because someone at Cal Tech had started first.[17]

The Chicago-born Watson managed to push Crick into poaching on Wilkins's territory, and together the two men arranged to get hold of the most recent high-quality X-ray portraits of DNA—painstakingly difficult work done by Wilkins's assistant Rosalind Franklin, a skilled scientist in her own right who was consistently dismissed by the old boys at Cambridge as no more than a troublemaking "bluestocking." ("The best home for a feminist," Watson remarked, in one of his less charming moments, "was in another person's lab.")

Meanwhile, the almost mythically famous American biochemist Linus Pauling had also turned his attention to DNA. Determined to one-up Wilkins and to "beat [Pauling] at his own game," Crick and Watson submerged themselves in DNA research. This was, as Watson writes, a question of asking "which atoms like to sit next to each other" and then building hypothetical structures, using "a set of molecular models superficially resembling the toys of preschool children." Their goal was to come up with a model that was consistent both with the X-ray pictures taken by Franklin and with all known chemical properties of the molecules involved.[18]

Early in 1953, Pauling proposed a possible solution: DNA was a "three-chain helix" with a sugar-phosphate "backbone" on the inside. Watson immediately realized that the model wouldn't work; the phosphate molecules would repel each other (or, as he put it, "Linus' chemistry was screwy"). But Pauling was clearly closing in on a workable model.

Watson threw himself into devising new models with even more energy ("I was racing [him] for the Nobel prize"). Obsessively doodling possible structures on reams of paper, building models from stiff cardboard, Watson finally settled on a new structure for DNA: a *double* helix, with the "backbone" on the outside. He and Crick calculated the chemical properties of this model, compared the existing X-ray data with the patterns that their model would produce, and decided that the hypothesis was sound. In April of 1953, they proposed this model in a short article published in the journal *Nature*; it concluded with a brief sentence (composed by Crick) suggesting that the double helix would allow nucleic acids to form hydrogen bonds—which meant DNA could be replicated. "It has not escaped our notice," Crick wrote, "that the specific pairing we have postulated immediately suggests a possible copying mechanism for the genetic material."[19]

The copying mechanism involved both double-stranded DNA and single-stranded ribonucleic acid (RNA); as biologist Colin Tudge explains, Crick and Watson envisaged the double strands of DNA splitting into

> two single strands, followed by the replication of each one. . . . A strand of DNA, once separated from its partner, can *either* begin to make a complementary copy of itself, and so replicate, *or* can begin to make a complementary strand of RNA, which then

leaves the nucleus and supervises the manufacture of appropriate protein in the cytoplasm.[20]

RNA served as the intermediary between the DNA and the newly manufactured proteins, but the exact way in which this happened (later researchers would eventually identify three different kinds of RNA with three different functions) was still unclear.

Even in its broader outline, though, the model was convincing: chemically sound, consistent with observed properties of nucleic acids. It was tested worldwide, elaborated upon by such biochemical luminaries as Frederick Sanger, George Gamow, Marshall Nirenberg, and J. Heinrich Matthaei. By the time James Watson published *The Double Helix: A Personal Account of the Discovery of the Structure of DNA* in 1968, the double-helical structure of DNA and its role in reproducing genetic material was accepted as gospel.

Francis Crick would later refer to the flow of information, from DNA to DNA, from DNA to RNA to protein, as the "central dogma" of modern biology, a label still widely used. Despite his use of the word "dogma," Crick knew that the theory was still speculative: "a grand hypothesis," he himself wrote, "that, however plausible, had little direct experimental support." In fact, the experimental support would not come along for another two decades. Not until the late 1970s would scientists have the technical tools to produce a truly detailed map of DNA, and protein-DNA interactions were not glimpsed until 1984. Watson's energetic and entertaining account is snappily titled, but neither he nor Crick had "discovered" DNA. Like Copernicus, they had instead built a convincing theory that accounted, very neatly, for decades of observable phenomena.[21]

JAMES D. WATSON
*The Double Helix: A Personal Account
of the Discovery of the Structure of DNA*
(1968)

Watson's original text is available as both a paperback reprint and an e-book. A more elaborate edition, containing editorial annotations, historical background, excerpts from personal letters, and additional illustrations, is available in hardcover and as an e-book from Simon & Schuster.

James D. Watson, *The Double Helix: A Personal Account of the Discovery of the Structure of DNA*, Touchstone (paperback and e-book, 2001, ISBN 978-0743216302).

James D. Watson, *The Annotated and Illustrated Double Helix*, ed. Alexander Gann and Jan Witkowski, Simon & Schuster (hardcover and e-book, 2012, ISBN 978-1476715490).

TWENTY-FOUR

Biology and Destiny

*The rise of neo-Darwinist reductionism,
and the resistance to it*

We are survival machines—robot vehicles blindly pro-
grammed to preserve the selfish molecules known as genes.
　　　　　—Richard Dawkins, *The Selfish Gene*, 1976

We are biological and our souls cannot fly free.
　　　　　—E. O. Wilson, *On Human Nature*, 1978

We are inextricably part of nature, but human uniqueness is
not negated thereby.
　　　　　—Stephen Jay Gould, *The Mismeasure of Man*, 1981

Watson and Crick's model carried all before it.

Continuing elaborations of the double helix only strengthened
the assumption that *this* was the missing piece of Darwin's puzzle:
the mechanism of inheritance, the location and engine of Mendel's
gene, the cornerstone of life. Much remained to be studied, but by
the 1960s the greatest mysteries of life seemed to have been solved.
"In the coiled structure of the DNA molecule and the complex
arrangement of its atoms lie the final secrets of heredity," marveled
Life magazine in October of 1965. "Scientists have begun to be
able to read the genetic code. . . . Once we can read, we may then
learn to 'write'—i.e., to give genetic instructions—in the DNA
code. When that time comes, man's powers will be truly godlike."
Godlike, that is, in the ability to create: "creatures never before

seen or imagined in the universe," or "new forms of humanity . . .
better adapted to survive on the surface of Jupiter, or on the bottom of the Atlantic Ocean." Or even, more simply, *ideal* human
beings: through the manipulation of DNA, "emphasizing man's
good qualities and eliminating the bad ones."[1]

Almost imperceptibly, DNA was being granted the power to
determine *who* we are, rather than merely *what* we are. And this
impulse was carried along by a field of study that existed even before
the Watson–Crick model sprang into being: population genetics.

Population genetics was intimately related to the "modern synthesis" of the 1930s and 1940s, which had attempted to reconcile
cell-level research with the larger story of organic life, the history
of the entire species with individual discoveries in microbiology.
In the same decade that T. H. Morgan and his team were mapping
fruit fly genes, the English biologist and statistician Ronald Fisher
was examining the odds that those genes would emerge in any
given generation. Fisher—a twin whose brother had died shortly
after birth, a Cambridge-trained mathematician who was more
interested in the life sciences than in his own field of study—is
often credited, along with Julian Huxley, Ernst Mayr, the British
mathematician and biologist J. B. S. Haldane, and a few others, as
a creator of the modern synthesis. In the 1930s, Fisher championed
the idea that the probability of certain genetic information passing
from one parent to one child could be calculated, and that these
calculations could be crunched into an even more basic conclusion:
a number predicting which individuals might survive, and which
might die.[2]

This was a paradigm-shifting idea. A century earlier, the "will
to live" had been simply accepted as part of what it meant to be
human. It was an ineffable, incalculable factor, a survival "instinct"
that could not be reduced to physical factors (actually, not so different from the present day).

But now, Fisher was proposing a *quantitative* definition of the
will to live—one that had a chemical basis, ungrounded in the free
human psyche. The will to live ("fitness," in Darwinian terms)
had to do with body instead of spirit, organs instead of determination. It depended on a certain combination of genes, which

Fisher envisioned as a bag of factors that behaved not unlike gas molecules in a chemical laboratory: randomly mixing, mating, and replicating.

That mixture of genes might lead to a longer claw, a more powerful muscle, better camouflage coloring, or a characteristic refusal to lie down and die. But all four were the product of gene mixtures; all four could be predicted, should your mathematics be good enough. Courage, like a sharp tooth, was a genetic characteristic—not a transcendent quality. And good math meant that you could predict how often (and, very roughly, where) courage, or sharp teeth, would show up in later generations. Population genetics attempted to do just this: to calculate when genetic variants would show up as changes in appearance or behavior, and then to compare those predictions with the characteristics of a current population of plants, or animals, or (even) people.[3]

Population genetics was a tricky, squeaky, infant field. Its early practitioners worked on the assumption that *alleles* (particular forms of a gene, located at specific places on the chromosome) could be passed from parent to child; the round shape of a pea was determined by an allele, as was an oblong and wrinkled shape. Where these characteristics might show up was mostly, although not entirely, a mathematical calculation: "At present," J. B. S. Haldane wrote in 1938, "one may say that the mathematical theory of evolution is in a somewhat unfortunate position, too mathematical to interest most biologists and not sufficiently mathematical to interest most mathematicians. Nevertheless, it is reasonable to suppose that in the next half century it will be developed into a respectable branch of Applied Mathematics."[4]

Haldane himself was already committed to the mathematical theory of evolution. Even before knowing exactly how genetic information was transmitted to the next generation, he had theorized that our will to survive was related to the urge to preserve our particular alleles. Famously, when asked whether he would be willing to sacrifice himself for a brother, he gave a mathematician's answer: "No," he reputedly replied. "Two brothers or eight cousins." Both would preserve the same amount of genetic material.[5]

After publication of the Watson-Crick model, that "genetic

material" was more clearly understood: the will to survive was related to the desire to protect, and pass on, a particular arrangement of DNA. In the early 1960s the English biologist William D. Hamilton proposed that DNA preservation could explain *unselfish* behavior, such as self-sacrifice. His hypothesis, nicknamed "Hamilton's Rule," suggested that altruism is always found in closely related organisms; when one creature sacrifices itself for another, it is always because they share genes. When a mother bird leads a cat away from her nest, at risk to her own life, the real beneficiary of this self-sacrificing behavior isn't the young birds; it is the *mother's* DNA, which has already been reproduced in her babies and now must be protected.

Hamilton pointed out that this theory explained an odd behavior in bees and ants: females in both species will sacrifice their babies to protect their sisters. Bees and ants have a particular kind of reproduction called *haplodiploidy*. Some of their eggs mature into adults without being fertilized by males; these eggs (oddly enough) produce male offspring. Eggs that *are* fertilized by mature males hatch into *females*. As a result, males have only one set of chromosomes, while females have two—which means that female bees and ants have more genes in common with their sisters than with their own offspring.[6]

This, Hamilton concluded, was why an ant will share food with her sisters and allow her children to starve. The probability of unselfishness, and its potential beneficiaries, could be calculated from the amount of DNA that two living creatures shared: the mathematics of altruism.

A few years later, the American chemist and geneticist George Robert Price proposed an equation that could predict such behavior: Price's Equation, an expression of the relationship between the genetic material of a parent and the genetic material in successful offspring. The equation,

$$\bar{w}\Delta\bar{z} = Cov(w_i, z_i) + E(w_i \Delta z_i)$$

where w stands for the fitness of an organism, and z for a particular attribute, can be used to predict the appearance, or absence, of *any measurable trait*: "an exact, complete description of evolution-ary change under all conditions," as evolutionary biologist Ste-ven A. Frank puts it. (Right after publishing this equation, Price converted to Christianity and gave away most of his goods to the homeless and poor; five years later he killed himself.)[7]

In 1976 the Oxford biologist Richard Dawkins published *The Selfish Gene*, a book that incorporated Hamilton's Rule, Price's Equation, and insights from population genetics into a macroex-planation: a comprehensive scientific explanation for all organic life, including ours. "Intelligent life on a planet comes of age when it first works out the reason for its own existence," Dawkins begins, and the reason he has worked out is a simple one: we eat, sleep, have sex, think, write, build space vehicles and war machines, sac-rifice ourselves or others, all in order to preserve our DNA. Natu-ral selection happens at the most basic level, the molecular; our bodies have evolved to do nothing more than protect and propa-gate our genes, which are ruthlessly selfish molecules working to ensure their own survival.[8]

It was not a comforting view of the world, and *The Selfish Gene* roused quite a bit of public furor. But Dawkins's conclusions were simply the logical outworking of the implications of Darwin-ian natural selection, combined with insights from the previous decades of population genetics and microbiology.

He had certainly not "invented the notion . . . that the body is merely an evolutionary vehicle for the gene" (as one science book claims), any more than Watson and Crick had "discov-ered" DNA. In fact, in 1975, the year before *The Selfish Gene* was published, the biologist E. O. Wilson had concluded (in the first chapter of his text *Sociobiology*) that "the organism is only DNA's way of making more DNA." But Dawkins was a good writer and a capable rhetorician, and *The Selfish Gene* managed to spell out the implications of this idea with particular clarity, accessible both to lay readers and to students of the life sciences. In the words of evolutionary biologist Andrew Read, a doctoral candidate when the book came out, "The intellectual framework had already been in the air, but *The Selfish Gene* crystallized it and made it impossible to ignore."[9]

•

The American biologist E. O. Wilson was hard on Dawkins's heels.

Twenty years earlier, Wilson had earned his reputation by demonstrating that the complex and sophisticated behavior patterns of fire ants were, in fact, based on chemical signals. Entomologists had struggled to explain *how* fire ants managed to communicate clearly with each other (were they tapping their antennae? stroking each other's bodies? emitting some other signal?). One possible explanation was that the ants were giving off chemical cues, but no one had identified *how*. Wilson, just turned thirty and fresh from his PhD, was trying all sorts of odd things with fire ants ("uncontrolled experiments," he called them much later, "quick and sloppy . . . performed just to see if you can make something interesting happen"). He tried to turn a line of marching ants with a powerful magnet. ("The ants couldn't care less.") He tried chilling colonies and switching their queens to see if he could mix different species together; this actually worked. And he picked out all of the major organs in the abdomen of worker ants, most of the organs no thicker than a single thread, and tried to identify one of them as the emitter of those theoretical chemical signals.[10]

One of the organs, the "almost invisible" Dufour's gland, just above the sting, came up positive. When Wilson used the dissected gland to mark a trail, other fire ants poured out of their nest to follow it. Wilson had proved the existence of pheromones, chemical substances that have the power not merely to direct, but also to intensify, galvanize, *change* behavior.

Behavior rests on chemistry: this became one of the cornerstones of Wilson's approach to organic life. His evolving philosophy was *disciplinary reductionism*; insights from physics and chemistry, demonstrable through experimentation, able to be confirmed by calculation, were the bedrock of all human knowledge. Biology rests on this bedrock; biological laws are directly derived from physical and chemical principles. And the social sciences—psychology, anthropology, ethology (natural animal behavior), sociology—float above, entirely dependent upon the "hard" sciences beneath.[11]

In the early 1960s, Wilson encountered Hamilton's work on the preservation of DNA in related organisms, a theory that had now been labeled "kin selection": "I became enchanted," Wilson later wrote, "by the originality and promised explanatory power of kin selection." As he continued to study insect behavior, he—like Dawkins—synthesized biochemical discoveries, population genetics, kin selection, and his own field observations into two texts that became classics. *The Insect Societies* (1971) argued that ant colonies behave like organisms, dividing labor, individuals sacrificing themselves for the whole, each ant less an individual than a part of the whole, a cog in the machine; the behavior of this entire society could be explained, predicted, entirely accounted for by physical and chemical factors.[12]

And then, in 1975, Wilson published *Sociobiology: The New Synthesis*, which extended the same argument to *all* societies, including ours. Human behavior, no less than ant actions, resulted from nothing more transcendent than physical necessity. Even seemingly intangible feelings and motivations (hate, love, guilt, fear) are

> constrained and shaped by the emotional control centers in the hypothalamus and limbic system of the brain. . . . What, we are then compelled to ask, made the hypothalamus and limbic system? They evolved by natural selection. . . . The hypothalamus and limbic system are engineered to perpetuate DNA.[13]

We are flooded with remorse, or the impulse to altruism, or despair, only because our brains (independent of our conscious knowledge) are reacting to our environment in the way that will best preserve our genes.

"Sociobiology," then, was the attempt to understand human society solely as a product of biological impulse. Ethics, philosophy, sociology, psychology: Wilson predicted that all would give way to *real* science, which, in the last analysis, boiled down to molecular biology.

> The conventional wisdom also speaks of ethology, which is the naturalistic study of whole patterns of animal behavior, and its

companion enterprise, comparative psychology, as the central, unifying fields of behavioral biology. They are not; both are destined to be cannibalized by neurophysiology and sensory physiology from one end and sociobiology and behavioral ecology from the other. . . . The future, it seems clear, cannot be with the ad hoc terminology [and] crude models . . . that characterize most of contemporary ethology and comparative psychology. Whole patterns of animal behavior will inevitably be explained within the framework, first, of integrative neurophysiology, which classifies neurons and reconstructs their circuitry, and, second, of sensory physiology, which seeks to characterize the cellular transducers at the molecular level.[14]

"I hope not too many scholars of ethology and psychology will be offended by this vision," Wilson added, which was unreasonably optimistic. Social scientists, as well as a healthy number of biologists, reacted with predictable outrage. In the fall of 1975, a group of scientists that included Wilson's Harvard colleague Stephen Jay Gould formed an opposition organization called the "Sociobiology Study Group," and published its objections in an open letter in the *New York Review of Books*. Wilson's sociobiology, they pointed out, was *deterministic*: it removed free will and choice from human society, making our current state seem inevitable.

Determinist theories . . . consistently tend to provide a genetic justification of the status quo and of existing privileges for certain groups according to class, race or sex. Historically, powerful countries or ruling groups within them have drawn support for the maintenance or extension of their power from these products of the scientific community. . . . These theories provided an important basis for the enactment of sterilization laws and restrictive immigration laws by the United States between 1910 and 1930 and also for the eugenics policies which led to the establishment of gas chambers in Nazi Germany.

Wilson's *Sociobiology*, the writers protested, was simply another version of these same dangerous ideas.

[Its] supposedly objective, scientific approach in reality conceals political assumptions. Thus, we are presented with yet another defense of the status quo as an inevitable consequence of "human nature." . . .

Wilson joins the long parade of biological determinists whose work has served to buttress the institutions of their society by exonerating them from responsibility for social problems. From what we have seen of the social and political impact of such theories in the past, we feel strongly that we should speak out against them.[15]

Wilson's response, which was to accuse his opponents of Marxism, didn't advance the discussion much.

But while the name-calling continued, other biologists, biochemists, and geneticists rallied behind *Sociobiology*. Over the next twenty years, Wilson's "new discipline" would give birth to yet another science: evolutionary psychology.

Wilson came to his own defense with *On Human Nature*. Published three years after *Sociobiology*, the book focused in, more closely, on humans. All of *Sociobiology* except for the last chapter had been based on animal research: "The final chapter," Wilson later mused, "should have been a book-length exposition . . . to address in a focused manner the main objections that had arisen and yet might arise from political ideology and religious belief. . . . [I wrote] *On Human Nature* in an attempt to achieve these various ends."[16]

On Human Nature did not back away from the "admittedly unappealing" conclusions of *Sociobiology*: "The human mind," Wilson began, "is a device for survival and reproduction, and reason is just one of its various techniques." He then explained how each of our most treasured attributes arise from our genes (so, for example, "The highest forms of religious practice . . . can be seen to confer biological advantage," not to mention that "genetic diversification, the ultimate function of sex, is served by the physical pleasure of the sex act"). And he closed his screed with a paean to scientific thinking:

[its] repeated triumphs in explaining and controlling the physical world; its self-correcting nature open to all [who are] competent to devise and conduct the tests; its readiness to examine all subjects sacred and profane; and now the possibility of explaining traditional religion by the mechanistic models of evolutionary biology. . . . In the end . . . the evolutionary epic is probably the best myth we will ever have.[17]

Like James Watson and Richard Dawkins, Wilson proved to be a talented writer, with a knack for powerful metaphors. *On Human Nature* was praised, excoriated, and read; it was an instant best seller, and in 1979 it won a Pulitzer Prize.

·

In 1981, Stephen Jay Gould struck back.

Gould, twelve years Wilson's junior, had already secured his place in the pantheon of evolutionary biologists by proposing (along with Niles Eldredge) a halfway place between uniformitarianism and catastrophism: punctuated equilibrium, the theory that species remain essentially the same for very long periods of time, interspersed with (relatively) fast periods of significant change. And, like Wilson, Gould was a good writer: a regular essayist for the general-interest magazine *Natural History*; the author of numerous professional works, as well as two highly popular science books for the general public.

His response to *On Human Nature*, titled *The Mismeasure of Man*, was (like Wilson's own book) aimed at a general readership. It was a focused and powerful refutation of *one* specific instance of biological determinism: the "abstraction of intelligence" as a biochemically determined quality, its "quantification" as a number (thanks to the increasing popularity of IQ tests), and "the use of these numbers to rank people" in a biologically determined "series of worthiness."

The argument was intended to play a much larger role than simply debunking IQ tests: Gould intended it to refute the biological determinism and disciplinary reductionism so prominent in Wilson's works. "*The Mismeasure of Man* is not fundamentally about the general moral turpitude of fallacious biological arguments in social settings," he wrote, in his introduction. "It is not even about the full range of phony arguments for the genetic basis of human inequalities" (a clear shot at *Sociobiology*). Rather,

> *The Mismeasure of Man* treats *one particular form* of *quantified* claim about the ranking of human groups: the argument that intelligence can be meaningfully abstracted as a single number capable of ranking all people on a linear scale of intrinsic and unalterable mental worth. Fortunately—and I made my decision on purpose—this limited subject embodies the deepest (and most common) philosophical error, with the most fundamental and far-ranging social impact, for the entire troubling subject of nature and nurture. . . . If I have learned one thing as a monthly essayist for over twenty years, I have come to understand the power of treating generalities by particulars.[18]

Like Wilson, Gould was assailed by some ("More factual errors per page than any book I have ever read," snapped the prominent psychologist Hans Eysenck, himself a believer in the genetic basis of intelligence) and praised by others (the book won the National Book Critics Circle Award in 1982). The battle lines—both of them—were drawn.[19]

And they remain, more or less, in the same place. In the twenty-first century, we still hear much about the struggle between evolutionary scientists and creationists (at least in the United States). But the struggle between biological determinists and evolutionary biologists who rejected determinism is far more wide-ranging and complicated. In 1997, Gould complained bitterly about what he called "Darwinian fundamentalism," the use of natural selection to explain *all* of life. To see human beings as merely genes "struggling for reproductive success" is, for Gould, "a hyper-Darwinian idea that I regard as a logically flawed and basically foolish caricature of Darwin's genuinely radical intent."[20]

Gould and his followers believed that other factors were at work—not divine intervention, but multiple, overlapping factors too complex to be reduced to simple survival of the gene. The

evolution of human intelligence was certainly one of these factors, but there were still more to be found. "We live in a world
of enormous complexity in organic design and diversity," Gould
concluded, shortly before his death from cancer in 2002,

> a world where some features of organisms evolved by an algorith-
> mic form of natural selections, some by an equally algorithmic
> theory of unselected neutrality, some by the vagaries of history's
> contingency, and some as by-products of other processes. Why
> should such a complex and various world yield to one narrowly
> construed cause? Let us have a cast . . . some more important and
> general, others for particular things—but all subject to scientific
> understanding, and all working together in a comprehensible way.[21]

RICHARD DAWKINS
The Selfish Gene
(1976)

The first edition can be easily located secondhand, but the thirtieth-
anniversary (third) edition, published in 2006, contains an updated
bibliography and a new introduction, as well as the prefaces from
the first and second editions.

Read the whole book, but note especially Chapter Nine, where
Dawkins discusses the ways in which cultural as well as biochemical information is transmitted from generation to generation.
Looking for a name for a "unit of cultural transmission" (he offers,
as examples, "tunes, ideas, catch-phrases, clothes fashions, ways
of making pots or of building arches"), Dawkins abbreviates the
Greek word *mimeme* to *meme*, thus contributing a brand-new (and
now common) word to the English language.

Richard Dawkins, *The Selfish Gene*, Oxford University Press
    (hardcover and paperback, 1976, ISBN 978-0198575191).
Richard Dawkins, *The Selfish Gene*, 30th anniversary edition,
    Oxford University Press (paperback and e-book, 2006, ISBN
    978-0199291151).

E. O. WILSON
On Human Nature
(1978)

Hardcover copies of the first edition are widely available. The 2004 revision contains a useful preface by Wilson, reflecting on the public reception of the original book.

Edward O. Wilson, *On Human Nature*, Harvard University Press (hardcover, 1978, ISBN 978-0674634411).

Edward O. Wilson, *On Human Nature*, revised edition (with a new preface), Harvard University Press (paperback and e-book, 2004, ISBN 978-0674016385).

STEPHEN JAY GOULD
The Mismeasure of Man
(1981)

Paperback copies of the 1981 edition can be located secondhand; W. W. Norton published a revised and expanded edition of the title, including Gould's updated defense of his argument and his interaction with biological determinism in the years since original publication.

Stephen Jay Gould, *The Mismeasure of Man*, W. W. Norton (paperback, 1981, ISBN 978-0393300567).

Stephen Jay Gould, *The Mismeasure of Man*, revised and expanded edition, W. W. Norton (paperback and e-book, 1996, ISBN 978-0393314250).

PART

V

READING
THE COSMOS

(*Reality*)

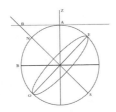

Albert Einstein, *Relativity:
The Special and General Theory* (1916)

Max Planck, "The Origin and Development
of the Quantum Theory" (1922)

Erwin Schrödinger, *What Is Life?* (1944)

[Edwin Hubble, *The Realm of the Nebulae* (1937)]

Fred Hoyle, *The Nature of the Universe* (1950)

Steven Weinberg, *The First Three Minutes:
A Modern View of the Origin of the Universe* (1977)

James Gleick, *Chaos* (1987)

TWENTY-FIVE

Relativity

The limits of Newtonian physics

We require to extend our ideas of the space-time continuum still farther.

—Albert Einstein, *Relativity:*
The Special and General Theory, 1916

For nearly two centuries, the universe was Newtonian.

His principles governed all investigations of the *cosmos*: all that there is, the solar system and the galaxy, the galaxies beyond and the stars, the earth and what is within and on it. The Newtonian cosmos was ruled by universal laws that always worked the same, in every place. ("If a property can be demonstrated to belong to all bodies on which experiments can be made," the third Rule for the Study of Natural Philosophy had declared, "it can be assumed to belong to all bodies in the universe.") Gravity functioned in the same way in every corner of the universe. Time passed, everywhere, at the same rate. Motion was *absolute*; it could, at least in theory, be measured according to some fixed, unchanging point in fixed, unchanging space ("absolute space"). The universe was static and infinite; it was neither expanding nor contracting (both would change the ways in which universal laws function), and it went on forever.

Even from the beginning, an occasional voice objected.

In 1721, just before the third edition of the *Principia* was published, the mathematician and philosopher George Berkeley questioned the existence of Newton's "absolute" space and time. Since man can measure space and time using only his own senses, Berke-ley argued, all motion has to be relative—it must be measured in relation to our own position, our own understanding. To propose the existence of *absolute* motion and space was to theorize beyond what science was capable of measuring; it was to edge over into the province of philosophy. "The philosopher of nature," he wrote, pointedly, in his essay *De motu*, "should remain entirely with his experiments, his laws of motion, his mechanical principles, and the conclusions derived therefore; if he has something to say on other matters, he should relate what is accepted in the respective higher science."

Newton, in other words, should stick to formulas, and leave questions of absolute existence to the "higher science" of philoso-phy. "Consider motion as something sensible . . . and . . . be con-tent with relative measures," Berkeley advised.[1]

But in practice, Newtonian physics triumphed—because it worked so extremely well.

In fact, it worked even better than Newton himself expected. His laws of gravity and motion made it possible to predict the movements of heavenly bodies with astonishing accuracy. But the gravitational forces of the solar system were so complicated, each body acting on another, constantly changing with motion, that it couldn't possibly run on its own indefinitely; Newton seems to have assumed that God would occasionally need to "reset" its delicate equilibrium. Certainly he believed that such a labyrinthine system demanded, at the very least, a divine send-off. "Though gravity might give the planets a motion of descent towards the sun," he wrote, in the early 1690s, "yet the transverse motions by which they revolve in their several orbits required the divine arm to impress them." And in another letter, "So then gravity may put the planets into motion, but without the divine power it could never put them into such a circulating motion as they have about the sun."[2]

A century after the *Principia* first appeared, the French math-ematician and astronomer Pierre-Simon Laplace began a twenty-five-year, five-volume set of calculations intended to dem-onstrate that Newtonian physics not only explained every motion in the solar system, but accounted for its eternal stability.

He succeeded. Much later, the story began to circulate that the Emperor Napoleon (whom Laplace had, very briefly, served as minister of the interior) had criticized the five volumes of the *Treatise on Celestial Mechanics* for never mentioning God. "Sir," Laplace is said to have answered, "I have no need of that hypothesis."[3]

Whether or not the conversation ever took place, Laplace's answer rings true. It was not a declaration of atheism, merely a statement of fact. The solar system had no need of a divine finger to tap it back into place.

In fact, Laplace also argued, in his less mathematical and more popular *System of the World*, that the divine finger was unnecessary at the very beginning as well. Newton's laws of gravitation suggested that the sun and planets could have coalesced from a rotating cloud of gas particles, each attracting the other until they clumped together into the bodies we now see. The philosopher Immanuel Kant had recently made a similar suggestion. It was very much a grand theory, immune to proof, but it fell in line with Laplace's commitment to Newtonian laws: They could explain the universe in toto. No other principle, no other explanation, was necessary.[4]

At least, until you got far away from the solar system, and into the deeper and more mysterious reaches of the universe.

·

Despite the third Rule, Newton never truly extended his beautifully calibrated mechanics to the farther corners of the cosmos. "The universe," for him, was the *known* universe: the stars, planets, and other objects that could be seen, tracked, plotted. He assumed that this universe was infinite, because in a finite universe, all stuff would eventually be pulled by gravity toward the center: in his own words, "the matter on the outside of this space would by its gravity tend towards all the matter on the inside, and by consequence fall down into the middle of the whole space, and there compose one great spherical mass." In an infinite universe, each particle would instead be pulled equally in all directions, producing equilibrium, *stasis*.[5]

Universal gravitation suggested that the individual masses (stars, planets, star clusters) should be more or less evenly distributed across this infinite space. But better and better telescopes revealed clumps, sparse spots, clusters.

Nor were these masses static. Fifty years after Laplace, the British astronomer William Huggins concluded that stars were *moving*, in relationship to the earth. The Austrian physicist Christian Doppler had recently shown that sound waves change in frequency when either the object emitting them or the hearer receiving them is moving; shortly afterward, the French researcher Armand Fizeau had extended this "Doppler effect" to the wavelengths of light. Huggins measured the changes in starlight (the "shift in . . . spectral lines") and used them to show that the star Sirius (among others) was receding from us, while others were approaching: "Speaking generally," he wrote, "the stars which the spectroscope shows to be moving from the earth . . . are situated in a part of the heavens opposite to Hercules . . . while the stars in the neighborhood of this region . . . show a motion of approach."[6]

In the Newtonian system, this movement was hard to account for.

At the same time that Huggins was discovering unexpected measurements, the mathematician Carl Friedrich Gauss was mounting a challenge to the foundation of *all* measurements: Euclidean geometry, which assumes that any stuff in the universe can be located within three dimensions (the x, y, and z coordinates: length, width, depth).* Intuitively, human beings resonate with Euclidean geometry, since we *live* in three dimensions. But Gauss had come to think that the infinite universe could not be expected to adhere only to methods easily understood by three-dimensional thinkers: "Finite man cannot claim to be able to regard the infinite as something to be grasped by means of ordinary methods of observation," he wrote to his colleague Heinrich Schumacher.[7]

Gauss played around with two-dimensional geometry and geometry done on a curve. (The curvature of a sphere, it turned out, could be calculated from a single point on the curve's surface, which meant that it wasn't necessary for three-dimensional space to *surround* the

---

*The mathematics of Gauss's challenge are well outside the scope of this book, but a useful explanation for the nonspecialist (complete with figures and illustrations) can be found in Eli Maor's *To Infinity and Beyond: A Cultural History of the Infinite* (Princeton University Press, 1991), 108–34.

point in order for it to curve—which was certainly anti-Euclidean.) But Gauss could not come up with a full alternative to Euclidean geometry that he could publish and stand behind: "Perhaps only in another life will we attain another insight into the nature of space, which is unattainable to us now," he wrote to a friend.[8]

Instead, Gauss challenged one of his students, the fragile and conscientious Bernhard Riemann, to pick up the challenge. Riemann became so immersed in the challenge that he suffered a nervous breakdown, but in 1854 he pulled himself together long enough to present his ideas to the faculty of the University of Göttingen.

There was, Riemann proposed, a *fourth* dimension. This dimension, which can be expressed algebraically, is impossible to visualize; it has to be explained through metaphor, as the theoretical physicist Michio Kaku does brilliantly:

Riemann imagined a race of two-dimensional creatures living on a sheet of paper. But the decisive break that he made was to put these bookworms on a *crumpled* sheet of paper. What would these bookworms think about their world? Riemann realized that they would conclude that their world was still perfectly flat. Because their bodies would also be crumpled, these bookworms would never notice that their world was distorted. However, Riemann argued that if these bookworms tried to move across the crumpled sheet of paper, they would feel a mysterious, unseen "force" that prevented them from moving in a straight line. They would be pushed left and right every time their bodies moved over a wrinkle on the sheet.[9]

Riemann then replaced the two-dimensional sheet with our three-dimensional world, crumpled in the fourth dimension. It would not be obvious to us that our universe was warped. However, we would immediately realize that something was amiss when we tried to walk in a straight line. We would walk like drunkards, as though an unseen force were tugging at us, pushing us left and right.

The existence of the fourth dimension suggested that neither Euclidean geometry *nor* Newtonian physics described the universe as it actually was. The fourth dimension implied that gravity and magnetism and electricity were not mysterious invisible "forces" that exerted power on objects. Instead they were geometric effects, caused by the warping of the fourth dimension.

This was a startling new way to look at the world: instantly appealing to a mathematician, because the calculations necessary to predict how these "forces" would work were potentially much simpler and more elegant than the calculations that Newtonian physics required. Working the calculations *out* was a massive task, however, and Riemann died of tuberculosis in 1866, aged thirty-nine, still struggling with the numbers.

Seven years later, the English mathematician William Clifford translated Riemann's space-shattering presentation to the Göttingen faculty; it was published for the first time in English in the journal *Nature*. Clifford suggested that the motion itself could also be explained simply by the distortion of the fourth dimension. "This variation of the curvature of space," he wrote, "is what really happens in that phenomenon which we call the *motion of matter.*"[10]

The mathematical exploration of this idea ran well ahead of the ability of physicists and astronomers to apply it to the real world, or to use it as an explanation of actual *phenomena*. But by 1900, more and more physicists were concluding that Newton's universe was on the skids; all that remained was to figure out the laws that governed the newly non-Euclidean universe.

.

In 1900, Albert Einstein had just received his first academic degree (from the Polytechnical Institute of Zurich) and, like most brand-new graduates, was finding the job market a tough one.

He had hoped to find a position doing advanced research, somewhere in the academic world. But no position appeared, and when a friend offered to help him get an interview at the Swiss patent office in Bern, he accepted. By 1902 he had become a Technical Expert Third Class—a patent examiner, in training.[11]

Einstein was well suited for the job, both by temperament (it was a quiet, solitary position that gave him plenty of time to think) and by training (his task was to evaluate patents involving electromagnetism, something he was particularly interested in). By 1905 he was in the running for a promotion to Technical Expert Second

Class and had drafted five different papers on various problems in electricity, magnetism, and related issues of space, time, and motion. One dealt with the movement of particles in liquid, others with the motions of atoms, the makeup of light. One of the papers proposed a formula for the conversion of energy into mass:

$$E = mc^2$$

where c equals the speed of light.

Einstein thought that the paper he completed on June 30, 1905, might be of special interest. Although the paper's title ("On the Electrodynamics of Moving Bodies") didn't suggest anything particularly revolutionary, it was (as Einstein wrote to a friend) nothing more than "a modification of the theory of space and time": Einstein's first exploration of what would later be known as the *special theory of relativity*.

The paper set out to reconcile two apparently contradictory laws. The first was the *principle of relativity*, which had been known since Galileo. A cornerstone of Enlightenment thinking, a classic Baconian assumption, the principle of relativity decrees that a law of physics must work in the same way across all related frames of reference. In his later version of the paper, intended for a general readership, Einstein used the example of a railway car, traveling along tracks next to an embankment, at a regular rate of speed. The car is constantly changing ("translating") its position relative to the embankment, but it doesn't simultaneously rotate, so this is called "uniform translation." At the same time, a raven is flying through the air, also in a straight line relative to the embankment, and also at a steady rate of speed—another uniform translation.

An observer standing on the embankment sees the raven flying at a certain rate of speed. An observer standing on the moving railway car, though, sees the raven flying at a *different* rate of speed.

If the embankment is called "coordinate system K" and the railroad car is "coordinate system K'," we can make this statement:

> If, relative to K, K' is a uniformly moving coordinate system devoid of rotation, then natural phenomena run their course with respect to K' according to exactly the same general laws with respect to K.[12]

25.1 EINSTEIN'S RAILWAY

In other words, for both observers the raven is still flying in the same direction and at a *constant* speed, even though the speed of the raven itself seems to be different.

Simple enough; but another law of physics contradicts the principle of relativity in a very fundamental way. "There is hardly a simpler law in physics," Einstein wrote, "than that . . . light is propagated in empty space . . . in straight lines with a velocity c = 300000 km./sec." The constant speed of light in a vacuum (air, water, and other transparent media slow it down) had been tested repeatedly since physicist Albert Michelson and chemist Edward Morley had accidentally discovered it in the early 1880s: "Let us assume," Einstein proposed, "that the simple law . . . is justifiably believed." What, then, was the problem?[13]

Imagine that a vacuum exists above the railway tracks, and that a ray of light travels above it, in the same direction as the raven. The principle of relativity insists that an observer on the embankment and an observer on the railway car will see the light traveling *at two different speeds*—which means that the speed of light is *not* constant.

What to do? It seems that either the principle of relativity or the constant speed of light needs to be abandoned; and, as Einstein pointed out, most physicists were inclined to abandon relativity ("in spite of the fact that no empirical data had been found which were contradictory to this principle"). But in fact, *neither* needs to be given up—as long as we are willing to adjust our ideas about time and space.[14]

The two observers were measuring the speed of light *per second*. Einstein suggested that what was changing was not the *speed* per second—but the second itself. Time, assumed to be a constant everywhere in the universe, was *not* constant at all. Time itself dilated, slowed down, as the observer moved faster. So the two observers were both measuring the speed per second of light, but for the observer who was *moving*, a second was *longer*. Time was the fourth dimension, the non-Euclidean addition; it turned three-dimensional Euclidean space into four-dimensional "space-time."

The special principle of relativity didn't take gravity into account (hence the adjective "special," or "limited"). But over the next ten years, Einstein struggled with gravity, looking for a theory of relativity that would incorporate gravitational pull.

By 1916 he had concluded that Bernhard Riemann was correct: gravity was a result, not a force. The presence of mass or energy (his formula allowed him to equate them) caused space-time (established by the 1905 special theory) to curve; objects traveling freely along the curves appeared to be *falling* but in fact were simply following "straight" along the surface of space-time. (Imagine that Riemann's bookworm existed, not on a crumpled sheet of paper, but on the surface of a rubber ball; the bookworm, unable to sense the curvature of its universe, crawls along the ball in what it thinks to be a straight line, but to an observer outside the curvature of the ball, the bookworm appears to be heading downward.)

The theory could be checked against effects caused by the sun, the most massive object nearby. Relativity explained an existing problem: the perihelion of Mercury, the point in its orbit that was closest to the sun, had *shifted*, or "precessed," over the previous centuries, and the precession was too large to be accounted for by the gravitational pull of the other planets. Einstein's new theory would account for it.

But there was a second test, one that would *predict* a phenomenon. If Einstein was correct, light from stars would be "pulled" toward the mass of the sun; starlight would be, observably, bent by the sun's mass.

Checking this theory required a total solar eclipse. *The General Theory of Relativity* was published in 1916, but Einstein's prediction was not confirmed until the British astronomer Arthur Eddington took measurements during a solar eclipse in 1919. Eddington's calculations showed that the starlight passing by the sun had shifted, to the exact degree that Einstein had foreseen.

The general theory of relativity told us that Baconian observation had its limits; that what we can *see* is not always what *is*; that common sense can lead us astray; that our senses can deceive us, although we'd better not ignore them. "Science is not just a collection of laws," Einstein wrote twenty years later. "It is a creation of the human mind, with its freely invented ideas and concepts. Physical theories try to form a picture of reality and to establish its connection with the wide world of sense impressions. Thus the only justification for our mental structures is whether and in what way our theories form such a link."[15]

Eddington's measurements had linked the "creation" of Einstein's mind, the general theory of relativity, to the world. Riemann's geometric theories had been put to use to describe actual sense impressions; physics had caught up to abstract mathematics, and had changed our picture of reality.

ALBERT EINSTEIN
The General Theory of Relativity
(1916)

Despite the equations, Einstein's paper is clear, elegantly written, and accessible even to nonmathematicians. The following edition is available in several formats, but Robert W. Lawson's 1920 translation into English is widely available; most editions include Einstein's summary of his findings on the special theory first. Read both, since the general theory builds on the special.

Albert Einstein, *Relativity: The Special and the General Theory*, trans. Robert W. Lawson, with introduction by Roger Penrose, commentary by Robert Geroch, and historical essay by David C. Cassidy, Pi Press (hardcover, paperback, and e-book, 2005, ISBN 978-0131862616).

TWENTY-SIX

Damn Quantum Jumps

The discovery of subatomic random swerves

The quantum . . . [will] play a fundamental role in physics, heralding the advent of a new state of things, destined . . . to transform completely our physical concepts.
— Max Planck, "The Origin and Development of the Quantum Theory," 1922

The great revelation of quantum theory was that features of discreteness were discovered in the Book of Nature, in a context in which anything other than continuity seemed to be absurd.
— Erwin Schrödinger, *What Is Life?*, 1944

Albert Einstein had a high capacity for new ideas. He could conceptualize the invisible bending of space; he could contemplate a space-time continuum that was quite unlike the three-dimensional reality in which he lived; he could make the imaginative leap into a world where time slowed to a standstill.

But he couldn't cope with quantum jumps. "I cannot seriously believe in it," he wrote to his friend Max Born, not long before Born won the Nobel Prize for his work in quantum mechanics. "The theory is incompatible with the principle that physics is to represent reality in time and space, without spookish long-distance effects."[1]

•

Those "spookish long-distance effects" were only one branch of the quantum physics that developed in the early twentieth century. But all quantum physics grew out of the same deep root: work done by chemists and physicists on the properties of atoms.

Those atoms had first been proposed by the Greek philosophers Leucippus and Democritus, who had suggested that all matter was made up of tiny particles, too small for the eye to see: *atomos*, the "undivided." Lacking proof, the hypothesis remained one among many. In the seventeenth century, Robert Boyle's experiments had suggested that the medieval version of atomic theory, a world constructed of corpuscles, was more likely true than not. But it would be another 150 years before the chemist John Dalton, building on experiments with gases done by many others (Joseph Black, Henry Cavendish, Joseph Priestley, and Antoine Lavoisier, to name a few), could restate the theory with conviction. Atoms, Dalton proposed, were indivisible; different atoms had different masses, and when only one type of atom was present, the matter in question was an element. Atoms of different types, mixed together, produced compounds.

Dalton's indivisible atom was an uncomplicated solid, but by the last quarter of the nineteenth century, a handful of physicists—among them, Joseph Thomson in Cambridge, Pieter Zeeman in Leiden, Walter Kaufmann in Bonn, and Emil Wiechert in Königsberg—were theorizing that the behavior of cathode rays (glowing beams observed when voltage was applied to vacuum tubes) could best be explained by the presence of smaller particles *within* atoms. Two Irish physicists, George Stoney and his nephew George Fitzgerald, were responsible for naming these smaller particles *electrons*: fundamental electrical units, carrying a negative charge.

This was not exactly the "discovery of the electron," as it is often described in textbooks. Atoms, as science philosopher Theodore Arabatzis points out, were not "observable entities" like bugs under rocks, so indisputable evidence of the existence of electrons, or atoms, was still missing. Rather, atomic theory was an effort to explain observable phenomena (like the bend in starlight as it passes the sun) by proposing underlying causes. But these proposals were very much hypothetical. They could be granted more or less weight, depending on how well mathematical models *based* on them predicted the behavior of observable physical phenomena. But they could not be proved—certainly not in any sense that Francis Bacon would have signed off on. In fact, Max Planck, the theoretical physicist who would later pioneer quantum theory, expressed his doubts about electrons; at the turn of the twentieth century, he still did not have "complete confidence in that theory."[2]

But over the next decade or so, calculations based on various aspects of atomic theory began to yield amazingly accurate results. In one of his 1905 papers ("On the Motion of Small Particles Suspended in Liquids at Rest, Required by the Molecular-Kinetic Theory of Heat"), Albert Einstein came up with a mathematical formula that predicted the properties of the apparently random movement of particles in water ("Brownian motion," first observed by Robert Brown in 1827) by relating them to the movements of those putative atoms.*

Einstein's calculations made it possible, theoretically, to estimate the number of atoms in a given substance, but his figures remained untested until 1908, when the French physicist Jean Perrin carried out two series of experiments that confirmed their validity. This research earned Perrin both the Nobel Prize and Einstein's thanks: "It is a piece of good luck for this subject that you undertook to study it," Einstein told Perrin the following year. It also convinced most physicists that the existence of the atom was no longer conjecture. "The atomic hypothesis has recently acquired enough credence to cease being a mere hypothesis," wrote another French physicist, Henri Poincaré. "Atoms are no longer just a use-

---

*For those interested in greater precision: Einstein's calculations were ostensibly about the motions of *molecules*, but his conclusions also affected two measurements, known as Avogadro's Number and Boltzmann's Constant, that were directly related to atomic theory. Avogadro's Number can be used to calculate the number of atoms in a given unit of a substance, and Boltzmann's Constant can predict the amount of thermal energy that those atoms carry. For a fuller explanation, see John S. Rigden, *Einstein 1905: The Standard of Greatness* (Harvard University Press, 2005), 57ff.

ful fiction; we can rightfully claim to see them, since we can actually count them."[3]

The next big question was the *structure* of the atom. Joseph Thomson had speculated (with no evidence whatsoever) that an atom was like a plum pudding, with the electrons ("corpuscles") sprinkled evenly throughout it. The problem was that, so far as he could see, electrons were all negatively charged, but atoms were electrically neutral.* "When they are assembled in a neutral atom," he wrote, visibly struggling with the missing piece of the puzzle, "the negative effect is balanced by *something* which causes the space through which the corpuscles are spread to act as if it had a charge of positive electricity equal in amount to the sum of the negative charges of the corpuscles."[4]

Jean Perrin guessed that there was another kind of particle within each atom:

> Each atom would consist, on one hand, of one or several positively charged masses—a kind of positive sun, the electric charge of which would greatly exceed that of a particle—and, on the other hand, by numerous particles acting as tiny negatively charged planets orbiting under the action of the electric forces, their negative total charge balancing exactly the total positive charge, thereby making the atom electrically neutral.[5]

Our solar system was a powerfully attractive metaphor; it made sense that Thomson's vague "something" might be a positive nucleus, orbited by the electrons. But Perrin acknowledged that this was only one of a number of possible models; it was merely a hypothesis, untested, unprovable.

A young German physicist named Hans Geiger came up with a way to look for the nucleus. Working with two colleagues—the distinguished Ernest Rutherford (last seen in Chapter 16, estimating the age of radioactive minerals) and a very young physics student named Ernst Marsden—he invented an instrument that could

---

*This wasn't quite right; atoms as a whole are not electrically neutral, but Thomson believed that they were, which led him to his next conclusion.

count the particles thrown off by decaying elements. This "Geiger counter" measured the amount of radiation being emitted, but Geiger and Marsden noticed something odd: if the particles were passed through various kinds of metal plates, they *changed direction* in a way that couldn't be accounted for by random motion. Some of them even went backward.

"It seems very surprising," Rutherford remarked, when reviewing these results. Something inside the atoms of the metal plates appeared to be colliding with the particles and bouncing them off into different trajectories. Thomson's "plum pudding" model suggested that particles should simply shoot right through atoms in their path, like buckshot passing through jelly; Rutherford concluded that an atom had to contain something more massive than an electron, something large enough to account for the deflection of the particles. This, he proposed in a 1911 paper, was "a central electric charge concentrated at a point and surrounded by a uniform spherical distribution of opposite electricity equal in amount." Working with this model, the "Rutherford atom," he was able to predict the movement of those pass-through particles—proof that each atom contained a nucleus, orbited by electrons.[6]

It was an elegant, intuitive model. It had a beautifully Platonic quality: the smallest particles in the universe mirroring the massive planetary movements in the heavens. Over a century later,

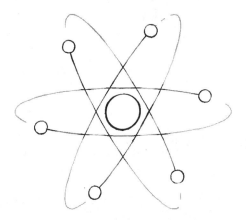

26.1 RUTHERFORD'S ATOM

the Rutherford atom is still the first picture that every chemistry student sees.

It turned out to be wrong. Sort of.

•

A decade before, the physicist Max Planck had been working on something called "blackbody radiation," radiation emitted by bodies that absorb the electromagnetic radiation that strikes them. (Hypothetically, a perfect blackbody object, *entirely* black, would suck in *all* the electromagnetic radiation coming its way.) In order to correctly predict the behavior of the radiation coming *out* of the blackbodies, Planck discovered that he had to fiddle with the properties of energy.[7]

According to every physical theory known, energy was a wave. It should be radiating out of those blackbodies smoothly, evenly, constantly. But Planck's calculations worked only if, instead, energy was pulsing out in *chunks*—not in waves, but in discrete units. If energy could be treated like separate particles (Planck called these hypothetical particles *quanta*, from the Latin *quantus*, "how much"), he could come up with a formula (now known as Planck's Constant) that explained the behavior of the radiation perfectly.*

Planck wasn't particularly happy about this solution. If a *quantum* were the size of a rock, it would be seen progressing forward in a series of jumps, not a smooth forward motion; this kind of movement seemed to contradict some of the most basic principles of physics and mechanics. Thirty years later, reflecting on his first formulation of these "quantum jumps," Planck wrote to a friend, "What I did can be described as simply an act of desperation. . . . It was clear to me that classical physics could offer no solution to this problem . . . [so] I was ready to sacrifice every one of my previous convictions about physical laws." But he pledged to keep looking for a more satisfactory solution; as far as he was concerned, quanta were "purely a formal assumption," a mathematical hat trick that

---

*A more technical but still readable explanation of Planck's investigations can be found in Bruce Rosenblum and Fred Kuttner, *Quantum Enigma: Physics Encounters Consciousness*, 2nd ed. (Oxford University Press, 2011), 55ff.

yielded the correct answers. Like scientists in the centuries before him, Planck was merely "saving the phenomena."[8]

Four years later, in one of his 1905 papers, Albert Einstein made use of Planck's work to explain some previously perplexing properties of light. "If we had to characterize the principal idea of quantum theory in one sentence," Einstein would later write, "we could say: *it must be assumed that some physical quantities so far regarded as continuous are composed of elementary quanta.*" Perhaps, Einstein theorized in 1905, light was *not* simply a continuous wave; perhaps it, too, was made up of individual particles.[9]

Like Planck, Einstein treated *quanta* as a theoretical construct, a technical fix rather than a picture of physical reality. But the more phenomena that quantum theory explained, the more "real" it seemed to be.

In 1913, the young Danish physicist Niels Bohr solved an atomic-level puzzle with the help of quantum theory. The puzzle had to do with the stability of the Rutherford model, which imagined electrons to be something like satellites circling the earth. If a satellite orbiting the earth lost some of its energy, it would spiral down and crash. When an atom emitted energy (as, for example, hydrogen atoms did, giving off light particles that some physicists had labeled *photons*), why did the orbits of the electrons not decay?

Bohr proposed that the orbits of electrons are *not* continuous circles. Rather, they are *quantized*: an electron does not sail smoothly through atomic space, but rather jumps from discrete spot to discrete spot. When a hydrogen atom emits a photon, the electron loses energy, but it doesn't spiral down; it "jumps" to a lower orbital path, one that is stable but takes less energy to maintain. The difference between the higher and lower orbits could be calculated. It corresponded *exactly* to the energy emitted by the atom.[10]

Einstein praised Bohr's results, but over the next decade or so, he and a whole cadre of other physicists became increasingly aware of just how odd the implications of quantum mechanics were. For example, in the new "Bohr–Rutherford model" of the atom, an electron could perform a "quantum leap" between orbits, rather than gliding smoothly through consecutive space. So, while *making* the leap, the electron was . . . nowhere.

And this was only one of the paradoxes that arose when quantum theory was used to solve existing physical problems. Quantum theory, announced Max Planck in his Nobel Prize address of 1922, had the potential "to transform completely our physical concepts," and this was a massive upset in the world of physics.[11]

•

In October of 1926, the Austrian physicist Erwin Schrödinger traveled to Bohr's home city of Copenhagen to discuss those upsets in person.

Schrödinger, two years Bohr's junior, was at the height of a distinguished academic career. He was in line to be Max Planck's successor as a lecturer in theoretical physics at Berlin University, and he had just published his own version of quantum theory—one that insisted on retaining waves to describe physical phenomena. Schrödinger appreciated the value of quantum theory in problem solving, but he was worried about doing away with waves. Without waves, an electron had no definite *path*, no trajectory through space; it could simply disappear and reappear, Cheshire Cat–like, with no way to predict the next place it would show up. Without waves, Schrödinger insisted, physics had no connection with reality, with the laws of electrodynamics, with our *experience*.

Bohr refused to yield the point. Quantum jumps had nothing to do with the everyday physics that governed ordinary life; they could not be directly experienced, but they were no less "real." According to Bohr's then-assistant, the young German physicist Werner Heisenberg, Schrödinger finally said, in exasperation, "If we are going to have to put up with these damn quantum jumps, I am sorry that I ever had anything to do with quantum theory."[12]

Schrödinger returned home, determined to resist the randomness and uncertainty of quantum jumps. Meanwhile, Heisenberg, still in Copenhagen, was figuring out just how those jumps could be better understood. The end point of a quantum jump, Heisenberg concluded, is impossible to calculate with certainty. We cannot predict exactly where a quantum particle will show up next; although we can calculate the *probability* of where it will reappear, we can only be certain of its new location once the particle has *actually* reappeared. But this, too, poses a problem: any instrument that is sensitive enough to measure the particle (for example, an electron microscope, which can locate particles by bouncing electrons off of them) will have to strike the reappearing particle, which will change its trajectory. In short, an exact measurement at a single point in time is, for all practical purposes, impossible. This conclusion, expressed mathematically, became known as the Heisenberg Uncertainty Principle.[13]

Heisenberg was quick to point out that, for objects larger than a molecule, our uncertainty is *minuscule*. Less than minuscule: essentially nonexistent. Only at the subatomic level does the uncertainty play any part in our understanding of the material world. An electron orbiting the nucleus of a hydrogen atom might make an unexpected leap, but a goat grazing on a hillside isn't going anywhere unpredictable at all.

This didn't reassure Schrödinger, who clung to the reality of *predictable* movement through space and time. His solution was an alternate quantum theory: *wave mechanics*. Wave mechanics turned Bohr's version of quantum theory on its head. What if, Schrödinger proposed, the movements of electrons were not because waves were actually particles—but because particles were actually waves? What if electrons themselves were merely the manifestation of a particular phase in a wave's existence?

Later, Albert Einstein would explain wave mechanics using the analogy of a rubber cord, shaken so that a wave travels down it:

> We take in our hand the end of a very long flexible rubber tube, or a very long spring, and try to move it rhythmically up and down, so that the end oscillates. Then . . . a wave is created by the oscillation which spreads through the tube with a certain velocity. . . .

26.2 EINSTEIN'S TUBE

. . .

Now another case. The two ends of the same tube are fastened.
. . . What happens now if a wave is created at one end of the rubber tube or cord? The wave begins its journey as in the previous
example, but it is soon reflected by the other end of the tube. We
now have two waves: one created by oscillation, the other by
reflection; they travel in opposite directions and interfere with
each other. It would not be difficult to trace the interference of
the two waves and discover the one wave resulting from their
superposition; it is called the *standing wave*.[14]

The standing wave had nodes, places where the waves canceled each
other out. Electrons, far from being discrete entities, moved along
the waveforms but were observable (*appearing* discrete) only at the
places farthest from the nodes—where the standing wave was greatest.

Mathematically, Schrödinger's wave mechanics and Bohr's
quantum leaps (which became known as the "Copenhagen
interpretation") actually ended up yielding very similar results.
The difference between them was, at base, a philosophical one.
Schrödinger's wave mechanics could not predict the position of an
electron, at any given time, with much more certainty than Bohr's

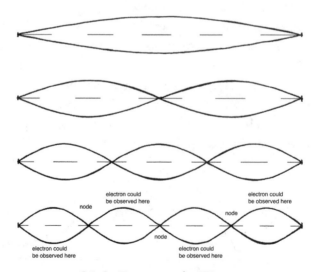

26.3 Einstein's Waves

Copenhagen theory; but Schrödinger had held on to a mechanical explanation that, while no more observable than quantum jumps, still resembled an effect that could be acted out in the real world. He had on his side Max Planck, and also Albert Einstein, who remained until the end of his life deeply skeptical of the Copenhagen interpretation, and wary of those "spookish long-distance effects" that had no discernible physical cause.

In fact, Einstein read, and approved of, Schrödinger's 1935 paper "The Present Situation in Quantum Mechanics," in which Schrödinger proposed a thought experiment intended to point out the invalidity of the Copenhagen interpretation.

> Imagine that a cat is penned up in a steel chamber, along with the following device (which must be secured against direct interference by the cat): in a Geiger counter there is a tiny bit of radioactive substance, so small, that perhaps in the course of the hour one of the atoms decays, but also, with equal probability, perhaps none; if it happens, the counter tube discharges and through a relay releases a hammer which shatters a small flask of hydrocyanic acid. If one has left this entire system to itself for an hour, one would say that the cat still lives if meanwhile no atom has decayed. The psi-function of the entire system would express this by having in it the living and dead cat (pardon the expression) mixed or smeared out in equal parts.

> It is typical of these cases that an indeterminacy originally restricted to the atomic domain becomes transformed into macroscopic indeterminacy, which can then be resolved by direct observation. That prevents us from so naively accepting as valid a "blurred model" for representing reality. In itself it would not embody anything unclear or contradictory. There is a difference between a shaky or out-of-focus photograph and a snapshot of clouds and fog banks.[15]

In other words, Schrödinger continued to believe that quantum theory could not simply be sequestered into a conveniently separate closet marked "subatomic." Whatever message it conveyed to us about the movement of electrons was the same message it was conveying about the rest of reality.

Einstein agreed. After reading an early draft of the paper, he wrote to Schrödinger, "Your cat shows that we are in complete agreement. . . . [A box that] contains the living as well as the dead cat just cannot be taken as a description of a real state of affairs."[16]

•

Thirty years earlier, there had been no such thing as quantum theory at all; now there were two vigorous, well-supported quantum theories, with plenty of adherents to both.

It was out of the conflict between the two theories that Schrödinger's 1944 book *What Is Life?* arose. In it, Schrödinger dealt with the overlap between quantum physics and biology, the common ground between the study of ourselves and the study of the cosmos. Using quantum theory to account for the behavior of orbiting electrons, Schrödinger showed how this behavior affected the formation of chemical bonds, and how those chemical bonds then affected cell behavior, genetics, evolutionary biology.

Like the works of Julian Huxley and Ernst Mayr, *What Is Life?* was a synthesis—an effort to show that quantum theory was *not* all about spookish, long-distance effects. Its success can be measured by the number of physicists who were inspired, after reading it, to migrate over into biological research. "No doubt molecular biology would have developed without *What Is Life?*" writes Schrödinger's biographer, Walter Moore, "but it would have been at a slower pace, and without some of its brightest stars. There is no other instance in the history of science in which a short semipopular book catalyzed the future development of a great field of research."[17]

"Semipopular" is an accurate description; *What Is Life?* was not a groundbreaking twentieth-century work of quantum theory. Pure quantum theory was mostly impenetrable to anyone except physicists. In fact, physics had been in danger of becoming a mountaintop pursuit, its discoveries expressed in a mathematical language so rarefied that only a few could keep up.

But for Schrödinger, quantum mechanics *had* to be able to

speak about the reality that is accessible to our senses. Sooner or later, quantum mechanics would affect the cat.

MAX PLANCK
"The Origin and Development of the Quantum Theory"
(1922)

Planck's brief essay, the written version of his Nobel Prize address, provides a fascinating glimpse into the development and early direction of quantum theory. The original English translation, by H. T. Clarke and L. Silberstein, is widely available online, as well as in paperback reprints.

Max Planck, "The Origin and Development of the Quantum Theory," trans. H. T. Clarke and L. Silberstein, Clarendon Press (e-book, 1922; paperback reprint available from Forgotten Books, 2013, ISBN 978-1440037849).

ERWIN SCHRÖDINGER
What Is Life?
(1944)

The standard edition is published by Cambridge University Press and includes Schrödinger's short essay on consciousness, "Mind & Matter."

Erwin Schrödinger, *What Is Life?: The Physical Aspect of the Living Cell; with, Mind & Matter; & Autobiographical Sketches*, Cambridge University Press (paperback and e-book, 1992, ISBN 978-1107604667).

TWENTY-SEVEN

The Triumph of the Big Bang

*Returning to the question of beginnings,
and contemplating the end*

The nebulae are all rushing away from our stellar system,
with velocities that increase directly with distances.
—Edwin Hubble, *The Realm of the Nebulae*, 1937

I find myself forced to assume that the nature of the
Universe requires continuous creation— the perpetual
bringing into being of new background material.
—Fred Hoyle, *The Nature of the Universe*, 1950

In the beginning, there was an explosion.
—Steven Weinberg, *The First Three Minutes:
A Modern View of the Origin of the Universe*, 1977

While quantum physicists peered closer and closer inward, astronomers extended their gaze outward to the skies.

Once again, Einstein's work helped to shape the ongoing investigations. If general relativity was a valid theory, three phenomena should be observable. First, the perihelion of Mercury should have shifted over time to a specific degree; Einstein himself had demonstrated this.* Second, starlight should bend as it passed the sun; Arthur Eddington had measured the bend in 1919.

---

* Over the next few years, his calculations were debated by other physicists but ultimately stood fast.

And third, there was the predicted redshift.

Forty years before Einstein, the British astronomer William Huggins had used wavelength changes in starlight to demonstrate that stars were moving, both toward and away from the earth—a visual version of the Doppler effect.* If Einstein's general theory of relativity was correct, astronomers should be able to detect a very *specific* kind of wavelength change in the light coming from massive bodies such as the sun. The mass curves space-time, meaning that light particles have to work *harder* to move away—as if the particles were scrambling up a smooth, curved wall, rather than skating away on a level surface. Because the particles are expending more energy (actually traveling a longer distance), their radiation shifts down to a lower frequency—put visually, toward the red end of the spectrum.

Detecting this "redshift" was not simple, as Einstein himself mused. "The spectral lines of sunlight, as compared with the corresponding spectral lines of terrestrial sources of light, must be somewhat displaced towards the red," he wrote, "[but] it is difficult to discover whether the inferred influence of the gravitational potential really exists." In the decades after 1916, numerous attempts were made to demonstrate redshift in the light coming both from the sun and from Venus, but scientists argued over the interpretation of the data. "The difficulty of deciding for or against the Einstein shift in the sun lies in the conflicting nature of the evidence itself," concluded the British astronomer John Evershed in 1919; redshift, after all, could be affected by pressure, temperature, or (as Huggins had demonstrated) movement.[1]

The problem of the redshift, and inconclusive data results, bedeviled Einstein's theory of relativity for years. But measuring redshift became a regular part of astronomical observations. And in the 1930s the astronomer Edwin Hubble was measuring redshift when his calculations led him to a new conclusion.

The universe was *much, much* larger than we had previously suspected. And it was not, as Albert Einstein himself had concluded, in a stable and static state.

*See Chapter 25.

•

Fifteen years earlier, a young Edwin Hubble had been finishing up his doctoral dissertation, a photographic study of the fuzzy celestial patches known as *nebulae* (from the Latin word for "mist"). Nebulae had been observed for centuries, but the limited range of telescopes made it impossible to see exactly where and what they were. Hubble had used a cutting-edge 24-inch reflecting telescope to photograph the nebulae, but although he was able to map their locations with some accuracy, he could only speculate about the nature of the luminescent clouds.

His speculation was a radical one: "Perhaps," Hubble wrote, "we see clusters of 'galaxies.'"[2]

Up until this point, scientists knew of only one galaxy—our own. The Milky Way had been named by the Greeks (*kyklos galaktikos*, "milky circle"), and since Galileo it had been known to contain uncounted stars. Little was known, though, about what might lie beyond its borders (wherever those might be). *Galaxy* might well be identical to *universe*.

Some astronomers had theorized that there might be other galaxies, or "island universes" floating outside our own, and Hubble's telescopic observations inclined him to think that nebulae might be those galaxies. But the photographs were inconclusive, and World War I interrupted his research. Just after receiving his doctorate, Hubble joined the US Army and headed overseas.

He spent the war in a division that never actually went into combat, and after the armistice he returned to his astronomical work. This time he managed to get a place at the Mount Wilson Observatory in California—an observatory that boasted a brand-new 100-inch telescope, the largest in the world. Here, for the first time, he was able to see *into* a nebula.

In 1923 he identified a very specific type of object inside the nebula known as M31: a Cepheid, a star that brightens and dims in a regular cycle. The astronomers Henrietta Leavitt (a Radcliffe graduate, and one of the few women pursuing astronomy in the early twentieth century) and Harlow Shapley (a natural enemy of Hubble's; the two men were professional rivals, and constitution-

ally inclined to loathe each other) had pioneered a method of measuring the distance to Cepheid stars.* Using their yardstick, Hubble
figured the distance to the Cepheid in M31 to be at least 930,000
light-years from the earth†—far, far larger than any previous estimate of the size of the Milky Way.

The implication was clear. Either the Milky Way was impos-
sibly large, or else the M31 nebula lay outside of our galaxy.[3]

Hubble's continuing observations of nebulae over the next ten
years led to his study *The Realm of the Nebulae* (1937), which presented his carefully researched and documented conclusions: not
only M31, but *forty-four thousand other galaxies* ("island universes")
were scattered throughout the cosmos. This was, in the words of
astronomers Jay Pasachoff and Alex Filippenko, a sort of Copernican revolution. Copernicus had startled us all by showing that the
earth is simply a planet orbiting a star, like many others; Hubble
showed that the Milky Way is "just one galaxy among the myriads
in the Universe." Never again would our galaxy be equated with
what there is. In the grand scheme of things, we had become that
much smaller.[4]

Which was not even the most startling conclusion Hubble
came to.

"Light which reaches us from the nebulae," he wrote in 1937,
using the term he continued to apply to galaxies, "is reddened in
proportion to the distance it has traveled. This phenomenon is . . .
evidence that the nebulae are all rushing away from our stellar
system, with velocities that increase directly with distances." The
redshifts did not necessarily confirm Einstein's general theory of
relativity (that confirmation, laboratory produced and *reproducible*,
would not come until 1959), but they revealed a somewhat unex-

---

*Measuring the distance to any given star sounds easy but wasn't. The historical
steps in developing the ability to measure celestial objects are interesting but off
topic; a more detailed but still nontechnical account can be found in Kitty Ferguson's *Measuring the Universe: Our Historic Quest to Chart the Horizons of Space and
Time* (Walker, 1999).
†Einstein had saved the speed of light as a constant, so a light-year (the distance
that light can travel over the course of a Julian year, 365.25 days) can be calculated
as 9,460,730,472,580.8 kilometers.

pected phenomenon. Those forty-four thousand galaxies were all *moving*—something that Hubble could measure, thanks to the red-shift in their light.[5]

Hubble's precise observations allowed him to come up with a formula for this expansion: Hubble's Constant, or H_o. Multiplying Hubble's Constant by the distance (*D*) of a galaxy produces the velocity (*V*) at which the galaxy is moving away:

$$V = H_o \times D$$

In other words, the farther away a galaxy has traveled, the faster it is moving.[6]

"This explanation," Hubble writes, "interprets red-shifts as Doppler effects, that is to say, as velocity-shifts, indicating actual motion of recession." He adds, cautiously, that further investigation is necessary:

> Nebular red-shifts . . . on a very large scale [are] quite new in our experience, and empirical confirmation of their provisional interpretation as familiar velocity-shifts, is highly desirable. Critical tests are possible at least in principle, since rapidly receding nebulae should appear fainter than stationary nebulae at the same distance. . . . The interpretation of red-shifts is [thus] at least partially within the range of empirical investigation.[7]

In other words, this was not the same sort of science that quantum physicists were doing; the redshift was *visible*, and better telescopes would make it even more visible. The "island universes" were, demonstrably, sailing through the cosmos.

The slightly more theoretical question was, *Why?*

On this point, Hubble was more cautious. "The explorations of space end on a note of uncertainty," he concludes in *The Realm of the Nebulae*. "With increasing distance [from the earth], our knowledge fades, and fades rapidly. Eventually, we reach the dim boundary—the utmost limits of our telescopes. There, we measure shadows, and we search among ghostly errors of measurement for landmarks that are scarcely more substantial." But, tentatively, he holds out an explanation: Galaxies are moving because the universe itself is expanding.[8]

This theory was not original with Hubble. In 1922 the young Russian physicist and mathematician Alexander Friedmann had suggested that Einstein's general theory of relativity required the universe to be expanding regularly outward.* Three years later, Friedmann died of typhoid, leaving his work unfinished; but in 1927 the Belgian astronomer Georges Lemaître came to the same conclusion.

Einstein had reviewed both papers, but he remained unconvinced. He believed that his theories made the most sense in a universe that contained a finite amount of matter and was *static*, remaining still and unchanged. If the universe appeared to be infinite, this was an illusion caused by the curvature of space and time. ("Your calculations are correct," he wrote to Lemaître, "but your physics is abominable.")[9]

But now, for the first time, visible proof could be brought to bear on the question. Einstein traveled to the Mount Wilson Observatory, met with Hubble, reviewed Hubble's results, peered through the 100-inch telescope—and changed his mind. "The redshift of distant nebulae has smashed my old construction like a hammer blow," he told a California audience that had gathered to hear him lecture.[10]

By the end of the 1930s, most astronomers and physicists agreed with Einstein: the universe was not static, but expanding. This had implications for its past form. Lemaître, following his calculations backward, had come to the logical conclusion that, if the universe is steadily expanding outward, at some point it must have been much smaller. In fact, far back at the *beginning* of the process, there must have been a "zero" hour when all matter in the universe was packed closely together into one point.

There was no way, under current physical laws, to explain this point, or to predict how it would behave, or even to understand *what* it was. So Lemaître borrowed a term from mathematics: it was a *singularity*, a place where the rules governing the world we know break down.[11]

---

*For a more detailed explanation, and the relevant equations, see the first two chapters of Helge Kragh, *Cosmology and Controversy: The Historical Development of Two Theories of the Universe* (Princeton University Press, 1996), 3ff.

This was not so much an explanation as the absence of one: reaching the point where he could no longer explain the universe, Lemaître had substituted the lack of explanation in its place, and given the blank spot a name. In 1931 he attempted to borrow from quantum theory to fill the gap. "If the world has begun with a single quantum," he wrote in the journal *Nature*,

> the notions of space and time would altogether fail to have any meaning at the beginning; they would only begin to have a sensible meaning when the original quantum had been divided into a sufficient number of quanta. If this suggestion is correct, the beginning of the world happened a little before the beginning of space and time.[12]

This got rid of the rabbit-in-the-hat singularity, but Lemaître's quantum, a "primeval atom" that somehow contained all matter now in the universe and *produced* space and time by beginning to expand outward, wasn't much easier to explain: "It may be difficult to follow up the idea in detail," Lemaître added, somewhat apologetically.

This proved to be an understatement. Nor was the singularity the only problem that needed solving. If the universe is expanding steadily outward (and has been doing so for quite a long time), shouldn't it be getting *thinner* at the center? Wouldn't a far-future observer find himself standing in the middle of empty space, since all of the receding galaxies would eventually be so far away that they could no longer be observed?

In 1948 the Yorkshire-born astronomer Fred Hoyle tackled the "thinning universe" problem and disposed of Lemaître's primeval atom at the same time. In ten billion years, he suggested, a theoretical observer will actually see just about the same number of galaxies that we see now, because

> new galaxies will have condensed out of the background material at just about the rate necessary to compensate for those that are being lost as a consequence of their passing beyond our observable universe. At first sight it might be thought that this could not go

on indefinitely because the material forming the background would ultimately become exhausted. The reason why this is not so, is that new material appears to compensate for the background material that is constantly being condensed into galaxies. . . . I find myself forced to assume that the nature of the Universe requires continuous creation—the perpetual bringing into being of new background material.[13]

Hoyle argued that this "continuous creation" model was more scientific than Lemaître's explanation. "The hypothesis that all matter in the universe was created in one big bang at a particular time in the remote past . . . [is] in conflict with the observational requirements," he insisted, in a widely heard 1949 radio talk. "This big bang hypothesis is . . . an irrational process that cannot be described in scientific terms."[14]

Hoyle—thirty-four years old, four years into his first academic post as a Cambridge lecturer in astrophysics—had a gift for rhetoric. Adherents to Lemaître's primeval-atom hypothesis were quick to point out that the proposed expansion was neither big (it started out *very* small, in fact) nor a bang (there was no explosion, just the steady expansion of space). But the term "Big Bang" stuck fast.

In 1950, Hoyle published his own "continuous creation" theory in a well-written, engaging book called *The Nature of the Universe*. It offered a coherent alternative to the Big Bang: Rather than emerging from a singularity or a (still undefined) quantum, the universe was in a "steady state." The universe had no beginning, and would have no end; tiny quantities of matter (only a few atoms per cubic mile) were constantly coming into being; the expansion of the universe, and the creation of matter between its existing particles, would continue forever.[15]

Clearly, this theory had its own difficulties. But, as Hoyle himself pointed out, they were no more fatal than the difficulties with Lemaître's hypothesis. It was no less likely that matter was being constantly created than that *all* matter had come into being at one time.

The Nature of the Universe garnered both an enormous popular following and widespread scientific support. More physicists and astronomers threw their weight behind the primeval-atom theory, but it was a smallish majority. Both proposals explained certain phenomena and failed to account for others; and in the end, both pulled an unexpected rabbit from an invisible hat. Matter *appeared.* Somehow.

But the steady-state theory had one great advantage: it didn't require the abandonment of all known physical laws at any point in the past.

.

In the absence of actual proofs, both primeval-atom supporters and steady-state thinkers inevitably ended up doing metaphysics: arguing origins and first principles, with no chapter and verse to cite.

The search for validation continued, though. Two years before *The Nature of the Universe* presented the steady-state theory to the general public, the Russian physicist George Gamow— a firm believer in the "big bang"—had coauthored a brief but well-regarded academic paper, proposing that the distribution of chemical elements in the observable universe could be accounted for only if the universe *had* expanded steadily outward from a primordial, superdense, molten singularity. Shortly afterward, one of Gamow's junior coauthors, Ralph Alpher, had taken the conclusions further and suggested that the singularity's enormous heat would have dissipated as the superdense starting point expanded.* But some of that heat would still be radiating around the universe—residual microwave radiation, a steady background presence in the cosmos, a detectable remnant of the "primeval fireball" (as a later researcher termed it, apparently having learned something from Hoyle's rhetorical flourishes).[16]

This cosmic background buzz would be a particular *type* of radiation. It would be on the "Planck spectrum"; its character would be determined by temperature alone, and the electromagnetic rays would be isotropic (the same in all directions), not emitted by any particular *body.* A steady-state system could not account for this type of radiation.

---

*A more detailed and quite readable account is found in Gamow's 1952 book *The Creation of the Universe,* intended for informed nonspecialists and reissued by Dover Publications in 2012 as an e-book.

If it even existed.

Cosmic background radiation remained theoretical until fifteen years after *The Nature of the Universe* appeared. In 1965 the Princeton astrophysicist Robert Dicke and his team were attempting to build a receiver that would be sensitive enough to locate cosmic background radiation; at the same time, 30 miles away, two physicists working for Bell Laboratories, Arno Penzias and Robert Wilson, kept picking up unexplained static in their state-of-the-art microwave antenna. Connected (fortuitously) by a mutual acquaintance, the Princeton and Bell Labs scientists analyzed the static. It was the right spectrum to be Gamow's leftover cosmic radiation; it was, as Gamow's theory implied, isotropic; it was the first pebble on the scale that would tilt cosmological theory toward the Big Bang.[17]

In the next few years, multiple measurements of this cosmic background radiation confirmed its existence, verified its temperature, and clarified its spectrum. The steady-state theory wilted faster than a morning glory. "It has always seemed remarkable to me," says the distinguished physicist and astronomer Woodruff Sullivan, who was finishing up his first science degree at the time, "that, for the great majority of the astronomical community in the 1960s, the steady-state theory was killed off . . . by a serendipitous discovery. . . . It died because of the fulfilment of a little-known, never emphasised prediction by its rival theory."[18]

Fred Hoyle continued to defend his steady-state universe. The existence of cosmic background radiation made him no more comfortable with that pesky singularity, a stubborn and violent violation of all observable laws of physics. He was not alone. Dennis Sciama, chair of physics at the University of Maryland, lamented the presence of cosmic background radiation in 1967: "I must add," he told a class of graduate students, "that for me the loss of the steady-state theory has been a cause of great sadness. The steady-state theory has a sweep and beauty that for some unaccountable reason the architect of the universe appears to have overlooked. The universe in fact is a botched job, but I suppose we shall have to make the best of it."[19]

By the end of the 1960s, physicists and astronomers had converted en masse (with the exception of Hoyle, who, twenty-five years later, was still adjusting the steady-state theory in an attempt to make it work). The Big Bang had triumphed: neither big nor a bang, sitting stubbornly at the beginning of the universe, refusing to obey its laws.

It took the general public a few more years to sign on. Cosmic background radiation was not an easy thing to understand. It was detectable only by highly specialized instruments, explicable only through advanced and obscure mathematics. The reasons for its presence were difficult to grasp. Steady-state theory had been so convincing, in part, because Hoyle was able to present it so well. The expansion of the universe from that superdense, hot singularity needed an equally able popularizer.

He appeared in 1977: Steven Weinberg, a theoretical physicist from New York who won the Nobel Prize two years later. Weinberg wrote *The First Three Minutes* at the same time that he was doing intensive and highly technical work on cosmic background radiation. But he was able to shift successfully from an academic to a popular voice. Like Hoyle, he had a gift for metaphor ("if some ill-advised giant were to wiggle the sun back and forth, we on earth would not feel the effect for eight minutes, the time required for a wave to travel at the speed of light from the sun to the earth"), and he was able to recast the science of the early universe as a gripping story. *The First Three Minutes* clearly lays out background information about the expansion of the universe, runs through the historical development of various explanations (including steady-state theory), and shows the necessity of cosmic background radiation.

It was the first widely read explanation of the Big Bang, and the catalyst for an explosion of books for lay readers on cosmology and theoretical physics over the next decade. The popular appetite for an origins story, created by Weinberg and fed by scores of other titles (John Gribbin's *In Search of the Big Bang: Quantum Physics and Cosmology*, Heinz Pagels's *Perfect Symmetry: The Search for the Beginning of Time*, James Trefil's *The Moment of Creation: Big Bang Physics from Before the First Millisecond to the Present Universe*, and many, many, *many* more) carved a path directly to Stephen Hawking's *A Brief History of Time*, which rode the swelling wave to outsell them all. ("Surely not another book on the Big Bang and all that stuff," physicist Paul Davies remembers thinking when he first saw Hawking's tome.)[20]

Yet, as groundbreaking as it was, *The First Three Minutes* shared the drawbacks of all origin stories. It demanded a leap of faith about the beginning of the universe; and it led, inevitably, to speculation about its end.

"There is an embarrassing vagueness about the very beginning," Weinberg wrote, in his introduction, "the first hundredth of a second or so." A later chapter, devoted to the problem of that first fraction of time, doesn't do much better:

> With the aid of a good deal of highly speculative theory, we have been able to extrapolate the history of the universe back in time to a moment of infinite density. But this leaves us unsatisfied. We naturally want to know what there was before this moment, before the universe began to expand and cool. . . . We may have to get used to the idea of an absolute zero of time—a moment in the past beyond which it is in principle impossible to trace any chain of cause and effect.[21]

Baconian science both had its limitations and was reluctant to admit them. Weinberg concedes that some aspects of the Big Bang resist explanation and at the same time struggles against them. "Embarrassing" implies that he *should* be able to answer all questions (Bacon would have agreed). It carves out little space for uncertainty.

So *The First Three Minutes* ends on a very certain note, even as it moves into speculation. In the absence of the continuously created material proposed by Hoyle, Weinberg writes, the universe must, ultimately, stop expanding; it will either simply cease, fading away into cold and darkness, or else "experience a kind of cosmic 'bounce,' and begin to re-expand. . . . We can imagine an endless cycle of expansion and contraction stretching into the infinite past, with no beginning whatever."

And then Weinberg moves without a break into metaphysics: "There is not much comfort in any of this," he writes,

> It is almost irresistible for humans to believe that we have some special relation to the universe, that human life is not just a more-or-less farcical outcome of a chain of accidents reaching

back to the first three minutes. . . . The more the universe seems comprehensible, the more it also seems pointless.[22]

Moving from singularity to meaning is irresistible, and also non-scientific. Comfort and despair both are, entirely, non-Baconian.

Although *The First Three Minutes* begins by contrasting an ancient creation myth (from the Norse *Edda*) to the scientific tales that Weinberg is about to unfold, the two stories actually bear a striking resemblance to each other. Weinberg's story of origins is complete with a projected apocalypse and (at the *very* end of the book) a moral, the goal of humanity:

> But if there is no solace in the fruits of our research, there is at least some consolation in the research itself. Men and women are not content to comfort themselves with tales of gods and giants, or to confine their thoughts to the daily affairs of life; they also build telescopes and satellites and accelerators, and sit at their desks for endless hours working out the meaning of the data they gather. The effort to understand the universe is one of the very few things that lifts human life a little above the level of farce, and gives it some of the grace of tragedy.[23]

Weinberg has moved deftly from physical questions, to the purpose of life: which is to glorify science, and to pursue it forever. The Baconian project has turned in on itself and swallowed its own tail; what began as the study of what could be verified by experiment has become a way to locate truths that can never be touched.

For the exceedingly persistent . . .

EDWIN HUBBLE
The Realm of the Nebulae
(1937)

The difficulty with Hubble's text is summed up by the 1983 retrospective review in the *New Scientist*, which exclaims, "A serious systematic account written for the general reader," and then concludes, "One suggests as a required first question in any PhD oral in astronomy,

'Have you read *The Realm of the Nebulae?*'" These are not, exactly, the same audiences, and Hubble's prose veers from the accessible to the impenetrable. However, the tenacious reader will find enough of interest to justify plowing through the Yale reprint version.

Edwin Hubble, *The Realm of the Nebulae* (Silliman Memorial Lectures Series), Yale University Press (paperback reprint edition, 2013, ISBN 978-0300187120).

For the rest of us . . .

FRED HOYLE
The Nature of the Universe
(1950)

Hoyle displays just how much authority an accessible, clear writing style lends to a scientific voice. *The Nature of the Universe* is out of print, but copies of the 1960 Harper hardcover are easily located secondhand. The most significant chapter of the text has been reproduced in the collection *Theories of the Universe: From Babylonian Myth to Modern Science,* edited by Milton K. Munitz (Free Press, 1965).

Fred Hoyle, *The Nature of the Universe,* HarperCollins (hardcover, 1960, ISBN 978-0060028206).

STEVEN WEINBERG
The First Three Minutes:
A Modern View of the Origin of the Universe
(1977)

The classic text, which has never gone out of print, was published in a second updated edition with a new foreword and an even more recent afterword (1993) by Basic Books.

Steven Weinberg, *The First Three Minutes: A Modern View of the Origin of the Universe,* Basic Books (paperback and e-book, 1993, ISBN 978-0465024377).

TWENTY-EIGHT

The Butterfly Effect

Complex systems, and the (present)
limits of our understanding

Where chaos begins, classical science ends.
—James Gleick, *Chaos*, 1987

The study of the universe, continually more rarefied, was now carried on at the highest levels mostly in equations. Cutting-edge theory was difficult even for readers with a firm grasp of calculus, essentially impenetrable to those who lacked it.

This inaccessibility gave great power to the popularizers, writers capable of translating academic shorthand into convincing narratives. Walter Alvarez was a geologist capable of turning a phrase; Dawkins, E. O. Wilson, and Gould were biologists who could write; Steven Weinberg, like Erwin Schrödinger, was both popularizer and physicist.

But in physics—particularly the branch of physics often labeled *cosmology*, the contemplation of the entire universe and its workings—the last major paradigm shift of the twentieth century was brought to light by an English major.

•

Even in the age of quantum mechanics and singularities, Newton's principles persisted.

For one thing, Newton's laws *work* so well in the everyday world. Drop your cookie on the floor, and Newton can tell you where it will land. Get on a train traveling 72 miles per hour, and Newton can tell you exactly when that train will intersect with the dump truck that's speeding toward you at an angle. Take off from Washington, DC, in a 747, and Newton can tell you exactly when you'll land in Paris (as long as the air traffic controllers cooperate).

In the nineteenth century, the French mathematician and astronomer Pierre-Simon Laplace* theorized that Newton's laws, derived from present conditions, could predict anything that might happen in the future. This version of determinism—the current state of the universe "completely defines its future"—led Laplace to conclude that an all-knowing, all-seeing but time-bound being could predict the future with absolute accuracy.

> An intellect which at a certain moment would know all forces that set nature in motion, and all positions of all items of which nature is composed, if this intellect were also vast enough to submit these data to analysis, it would embrace in a single formula the movements of the greatest bodies of the universe and those of the tiniest atom; for such an intellect nothing would be uncertain and the future just like the past would be present before its eyes.[1]

This theoretical intellect became known as "Laplace's Demon" (*demon*, in this context, meaning "a hypothetical entity" rather than "an evil spirit"). Laplace's Demon could not only see all forces in the natural universe and map where they were at any given time, but also had the ability to crunch the numbers and plot a future course from them. For Laplace's Demon, time was irrelevant; the universe looked the same running backward as forward. Theoretically, the Demon could calculate into the past with just as much accuracy as into the future.

The first scientist to suggest that this might not be as straightforward as Laplace argued was another Frenchman, the mathematician and physicist Henri Poincaré, around 1910. Working with systems that appeared to be quite simple, Poincaré ran into unexpected results. He could *not* always predict the outcome, even when the forces in question seemed straightforward; and he believed that tiny changes in the initial conditions, so slight that he could not easily detect them, were the cause. "Small differences in the initial

*See Chapter 25.

conditions produce very great ones in the final phenomena," he theorized. "A small error in the former will produce an enormous error in the latter. Prediction becomes impossible."[2]

For half a century, no one followed up on this insight.

.

In the 1960s, Laplace's Demon appeared on the scene, at least partially and in embryonic form: computers, which didn't have the ability to comprehend the forces in nature but were potentially able to crunch all that data in a way that no human mind could.

In 1961 the American mathematician Edward Lorenz was working on weather. He had always been fascinated by weather patterns, which most mathematicians ignored as too quixotic to map. Using brand-new computer technology, he had written code that should have taken various factors (wind distance and speed, air pressure, temperature, and so on) and used them to predict weather patterns.

One particular evening, Lorenz entered the numbers that represented his factors, and his computer program, obediently, predicted a pattern. Lorenz decided to double-check the pattern. He reentered the factors, but to save time he punched them in to only the third decimal place, instead of the sixth (as he had done the first time).

This should have made no difference at all, since the change in wind speed, or temperature, from the fourth decimal place on was absolutely insignificant—in fact, almost nonexistent. But to Lorenz's shock, the weather pattern that resulted began to diverge from the original . . . and then departed from it entirely. By the time it finished cycling, the program had produced an entirely *different* set of weather phenomena. In 1963 Lorenz published a paper chronicling these results: "Deterministic Nonperiodic Flow." Perhaps, he wrote, the weather system was *so* sensitive to those tiny starting shifts that minuscule changes *could* actually produce massively different results.[3]

This sensitivity to infinitesimal conditions could hardly have been calculated before the advent of computers and their ability to run multiple scenarios very quickly. But now, Lorenz's paper attracted a great deal of attention from other mathematicians, who also began using computers to solve these particular kinds of "nonlinear equations."* In 1972, Lorenz followed up his original paper with another, called "Predictability: Does the Flap of a Butterfly's Wings in Brazil Set Off a Tornado in Texas?" It was the first time that a butterfly's wing was used as an analogy for one of those tiny starting changes: the first use of the phrase "butterfly effect."[4]

In 1975, two other mathematicians, Tien-Yien Li and James A. Yorke, published a paper about nonlinear equations and their unpredictable results that, for the first time, gave this unpredictability a name. They called it *chaos*: an immensely powerful word for most English-speaking readers who, even in 1975, still knew something of its biblical use: utter formlessness, confusion, disorder.[5]

This was not exactly chaos in a mathematical sense. *Chaos* here means "unpredictability"—and not even ultimate, intrinsic unpredictability (as in, "No matter how much we know, we will not be able to predict the end result") but, instead, a contingent, *practical* unpredictability ("This system is so sensitive to microscopic changes in initial conditions that we are not, at the moment, capable of analyzing those initial conditions with the accuracy necessary to predict all possible outcomes").

A paper published the following year by the mathematically gifted biologist Robert May gave chaos theory a more realistic but less catchy name: "Simple Mathematical Models with Very Complicated Dynamics." May's paper extended these "chaotic" systems past weather into something a little more concrete: insect populations, with the proposal that *apparently* random fluctuations in the numbers of insects within a certain society could be understood as related to those initial conditions.

After May, studies in chaos theory and its application to various fields (physics, chemistry, biochemistry, biology) accelerated. But chaos theory was still in its early adolescence when the writer James Gleick—a *New York Times Magazine* columnist, freelance essayist, and science reporter—chose it as the subject of his first book.

Chaos: Making a New Science was (like *T. rex and the Crater of Doom*, like *The Double Helix* and *The Mismeasure of Man*) snap-

---

*Not all such equations have chaotic solutions; those that do are a subset of the larger category "nonlinear equations."

pily titled, beautifully written, and peppered with vivid metaphors. Like *T. rex* (which led directly, at least, to the movies *Deep Impact* and *Armageddon*, and indirectly to scores of others), like the Big Bang, chaos theory gripped the popular imagination. The "butterfly effect" became a household phrase, especially once Jeff Goldblum's rock-star scientist character in *Jurassic Park* gave worldwide audiences the shorthand version: "It simply deals with predictability in complex systems. . . . A butterfly can flap its wings in Peking, and in Central Park, you get rain instead of sunshine. . . . Tiny variations . . . never repeat, and vastly affect the outcome. That's unpredictability."

The word "chaos" is deceptive, though; especially to readers who get the theory without the equations. "Where chaos begins, classical science ends," Gleick writes, in his clear (and deservedly best-selling) account. But at its core, chaos theory turns out to be almost Newtonian. Laplace's Demon, with its vast resources of knowledge and unlimited computational ability, could—theoretically—have tracked the air, moving from that butterfly wing all the way through to its ultimate end in rainy Central Park.

We cannot predict the outcome of complex systems, in the end, not because they are unpredictable, but because we cannot yet see deeply enough into the factors that shape them. But, buried deep in chaos theory is the promise—justified or not—that this may not always be the case.

JAMES GLEICK
Chaos: Making a New Science
(1987)

Gleick's original 1987 text is still available secondhand; a slightly revised and updated second edition was published in 2008.

James Gleick, *Chaos: Making a New Science*, Viking (hardcover and paperback, 1987, ISBN 978-0670811786).
James Gleick, *Chaos: Making a New Science*, Penguin Books (paperback and e-book, 2008, ISBN 978-0143113454).

Notes

ONE The First Science Texts

1. Albert Einstein and Leopold Infeld, *The Evolution of Physics: The Growth of Ideas from Early Concepts to Relativity and Quanta* (Cambridge University Press, 1938), 33.
2. Robert Parker, *On Greek Religion* (Cornell University Press, 2011), xi, 6.
3. Malcolm Williams, *Science and Social Science: An Introduction* (Taylor & Francis, 2002), 10.
4. Francesca Rochberg, *The Heavenly Writing: Divination, Horoscopy, and Astronomy in Mesopotamian Culture* (Cambridge University Press, 2004), 226.
5. Aristotle, *Metaphysics* 1.3, in *Readings in Ancient Greek Philosophy: From Thales to Aristotle*, 4th ed., ed. S. Marc Cohen, Patricia Curd, and C. D. C. Reeve (Hackett, 2011), 2.
6. Plato, *Protagoras*, trans. Benjamin Jowett (Serenity, 2009), 25.
7. Plinio Prioreschi, *A History of Medicine*, vol. 1, *Primitive and Ancient Medicine*, 2nd ed. (Horatius Press, 1996), 42.
8. Hippocrates, "On the Sacred Disease," in *The Corpus: Hippocratic Writings* (Kaplan, 2008), 99.
9. Lawrence I. Conrad et al., *The Western Medical Tradition: 800 B.C.–1800 A.D.* (Cambridge University Press, 1995), 23–25; Pausanius, *Pausanias's Description of Greece*, trans. J. G. Frazer (Macmillan, 1898), 3:250; Hippocrates, *On Airs, Waters, and Places*, in *Corpus*, 117.

TWO Beyond Man

1. Lawrence I. Conrad et al., *The Western Medical Tradition: 800 B.C.–1800 A.D.* (Cambridge University Press, 1995), 23; Hippocrates, *On Ancient Medicine*, trans. Mark J. Schiefsky (Brill, 2005), 32.

2. Gerard Naddaf, *The Greek Concept of Nature* (SUNY Press, 1995), 1–2.
3. Aristotle, *Physics*, trans. Robin Waterfield, Oxford World's Classics (Oxford University Press, 1999), xi; Naddaf, *Greek Concept of Nature*, 7, 65–66; Aristotle, *The Metaphysics*, trans. William David Ross, in *The Works of Aristotle* (Franklin Library, 1982), 3:175.
4. Simplicius, *Commentary on the Physics* 28.4-15, quoted in Jonathan Barnes, *Early Greek Philosophy*, rev. ed (Penguin, 2002), 202; Aristotle, *On Democritus*, frag. 203, quoted in Barnes, *Early Greek Philosophy*, 206–7.
5. Steven Weinberg, *Dreams of a Final Theory: The Scientist's Search for the Ultimate Laws of Nature* (Vintage, 1994), 7–8; see also Chapter 27, "The Triumph of the Big Bang."
6. C. C. W. Taylor, *The Atomists: Leucippus and Democritus, Fragments* (University of Toronto Press, 1999), 214–15.
7. Naddaf, *Greek Concept of Nature*, 9.
8. George Sarton, *A History of Science: Ancient Science through the Golden Age of Greece* (Harvard University Press, 1964), 421–24; Benjamin Jowett, *The Dialogues of Plato in Four Volumes* (Charles Scribner's Sons, 1892), 2:458–59.
9. Plato, *The Dialogues of Plato*, trans. Benjamin Jowett (Hearst's International Library, 1914), 4:463.
10. Plato, *Plato's Timaeus: Translation, Glossary, Appendices, and Introductory Essay*, trans. Peter Kalkavage (Focus, 2001), 60–61.

THREE Change

1. George Sarton, *A History of Science: Ancient Science through the Golden Age of Greece* (Harvard University Press, 1964), 423.
2. Jennifer Vonk and Todd K. Shackelford, eds., *The Oxford Handbook of Comparative Evolutionary Psychology* (Oxford University Press, 2012), 42.
3. Sarton, *History of Science*, 539; Jonathan Barnes, ed., *The Cambridge Companion to Aristotle* (Cambridge University Press, 1995), 123–26.

FOUR Grains of Sand

1. Malcolm Williams, *Science and Social Science: An Introduction* (Taylor & Francis, 2002), 11; Lewis Wolpert, *The Unnatural Nature of Science* (Harvard University Press, 1992), 35–36; Keith Devlin, *The Language of Mathematics: Making the Invisible Visible* (W. H. Freeman, 2000), 20.
2. Kenneth S. Guthrie and David R. Fideler, *The Pythagorean Sourcebook and Library: An Anthology of Ancient Writings Which Relate to Pythagoras and Pythagorean Philosophy* (Phanes Press, 1987), 58.
3. Richard Mankiewicz, *The Story of Mathematics* (Princeton University Press, 2000), 24.
4. Guthrie and Fideler, *Pythagorean Sourcebook*, 60; Mankiewicz, *Story of Mathematics*, 24, 26; Devlin, *Language of Mathematics*, 21.

5. Scholium to Euclid's *Elements*, quoted in Richard J. Trudeau, *The Non-Euclidean Revolution* (Birkhäuser, 1987), 103.

6. Plato, *The Republic: The Complete and Unabridged Jowett Translation* (Vintage, 1991), 265, 279, 281.

7. Margaret J. Osler, *Reconfiguring the World: Nature, God, and Human Understanding from the Middle Ages to Early Modern Europe* (Johns Hopkins University Press, 2010), 13–14.

8. Plato, *Republic*, 280.

9. Guthrie and Fideler, *Pythagorean Sourcebook*, 178; Carl Huffman, *Archytas of Tarentum: Pythagorean, Philosopher and Mathematician King* (Cambridge University Press, 2005), 303–4; Aristotle, *Politics*, trans. Ernest Barker, Oxford World's Classics (Oxford University Press, 1988), 311.

10. Devlin, *Language of Mathematics*, 300.

11. Euclid, *The Thirteen Books of the Elements*, 2nd ed., trans. Thomas L. Heath (Cambridge University Press, 1908), 1.

12. Vitruvius Pollio, *Vitruvius: The Ten Books on Architecture*, trans. M. H. Morgan (Dover, 1960), 254; Mary Jaeger, *Archimedes and the Roman Imagination* (University of Michigan Press, 2008), 19.

13. Keith Kendig, *Sink or Float: Thought Problems in Math and Physics* (Mathematical Association of Virginia, 2008), 67.

14. Archimedes, "The Sand-Reckoner," in *The Works of Archimedes*, trans. Thomas. L. Heath (Cambridge University Press, 1897), 221–22.

15. Alan W. Hirshfeld, *Parallax: The Race to Measure the Cosmos* (Birkhäuser, 2000), 12, 14–15.

16. George Coyne and Michael Heller, *A Comprehensible Universe* (Springer, 2008), 22–24; Charles Seife, *Zero: The Biography of a Dangerous Idea* (Viking, 2000), pp. 51–52.

FIVE The Void

1. C. C. W. Taylor, *The Atomists: Leucippus and Democritus, Fragments* (University of Toronto Press, 1999), 60.

2. Epicurus, "Letter to Herodotus," in *Letters and Sayings of Epicurus*, trans. Odysseus Makridis (Barnes & Noble, 2005), 3–6.

3. Anthony Gottlieb, *The Dream of Reason: A History of Philosophy from the Greeks to the Renaissance* (W. W. Norton, 2000), 290, 303.

4. George Sarton, *A History of Science: Ancient Science through the Golden Age of Greece* (Harvard University Press, 1964), 495; Lucretius, *On the Nature of the Universe*, trans. Ronald Melville (Oxford University Press, 1997), xvii.

5. Lucretius, *On the Nature of the Universe*, rev. sub. ed., trans. Ronald E. Latham (Penguin Classics, 1994), 13–14.

6. Titus Lucretius Carus, *On the Nature of Things*, trans. John Selby Watson (Henry G. Bohn, 1851), 96.

SIX   The Earth-Centered Universe

1. K. P. Moesgaard, "Astronomy," in *Companion Encyclopedia of the History & Philosophy of the Mathematical Sciences*, ed. I. Grattan-Guinness (Routledge, 1994), 241–42; Margaret J. Osler, *Reconfiguring the World: Nature, God, and Human Understanding from the Middle Ages to Early Modern Europe* (Johns Hopkins University Press, 2010), 16.

2. Norriss S. Hetherington, *Cosmology: Historical, Literary, Philosophical, Religious, and Scientific Perspectives* (CRC Press, 1993), 74–76.

3. Osler, *Reconfiguring the World*, 15.

4. C. M. Linton, *From Eudoxus to Einstein: A History of Mathematical Astronomy* (Cambridge University Press, 2008), 48.

5. H. Floris Cohen, *How Modern Science Came into the World: Four Civilizations, One 17th-Century Breakthrough* (Amsterdam University Press, 2010), 53.

6. Moesgaard, "Astronomy," 243–45; Cohen, *How Modern Science Came*, 56.

7. David C. Lindberg, *The Beginnings of Western Science*, 2nd ed. (University of Chicago Press, 2007), 249.

8. Olaf Pedersen, *A Survey of the Almagest*, rev. ed. (Springer, 2011), 19; Linton, *From Eudoxus to Einstein*, 117; Albert van Helden, *Measuring the Universe: Cosmic Dimensions from Aristarchus to Halley* (University of Chicago Press, 1985), 171.

9. Lynn Thorndike, *A History of Magic and Experimental Science* (Columbia University Press, 1941), 5:332.

SEVEN   The Last Ancient Astronomer

1. H. Floris Cohen, *How Modern Science Came into the World: Four Civilizations, One 17th-Century Breakthrough* (Amsterdam University Press, 2010), 106.

2. "Preface," in *De revolutionibus*, quoted in Thomas S. Kuhn, *The Copernican Revolution: Planetary Astronomy in the Development of Western Thought* (Harvard University Press, 1957), 137.

3. Jack Repcheck, *Copernicus' Secret: How the Scientific Revolution Began* (Simon & Schuster, 2007), 48.

4. Nicolaus Copernicus, *Three Copernican Treatises*, trans. Edward Rosen (Dover, 1959), 57.

5. Ibid., 58–59.

6. Kuhn, *Copernican Revolution*, 140.

7. Cohen, *How Modern Science Came*, 106; C. M. Linton, *From Eudoxus to Einstein: A History of Mathematical Astronomy* (Cambridge University Press, 2008), 121, 126.

8. Maurice A. Finocchiaro, *Defending Copernicus and Galileo: Critical Reasoning in the Two Affairs* (Springer, 2010), xiv.

9. Quoted in Linton, *From Eudoxus to Einstein*, 126–27.

10. Wim Verbaal, Yanick Maes, and Jan Papy, eds., *Latinitas perennis*, vol. 1, *The Continuity of Latin Literature* (Brill, 2007), 133; Nicolaus Copernicus, *On the Revolutions of the Heavenly Spheres*, trans. Charles Glenn Wallis (Prometheus Books, 1995), 6.

11. Copernicus, *On the Revolutions*, 18.

EIGHT A New Proposal

1. Tycho Brahe, quoted in Joshua Gilder and Anne-Lee Gilder, *Heavenly Intrigue: Johannes Kepler, Tycho Brahe, and the Murder behind One of History's Greatest Scientific Discoveries* (Random House, 2004), 81.

2. Catherine Drinker Bowen, *Francis Bacon: The Temper of a Man* (Little, Brown, 1963), 100–102.

3. Brian Vickers, ed., *Francis Bacon: The Major Works* (Oxford University Press, 2002), xviii.

4. Francis Bacon, *The Philosophical Works of Francis Bacon in Five Volumes*, ed. James Spedding (Longman, 1861), 4:65.

5. Ibid., 81.

6. Jennifer Mensch, *Kant's Organicism: Epigenesis and the Development of Critical Philosophy* (University of Chicago Press, 2013), 147.

7. Bowen, *Francis Bacon*, 187.

8. Abraham Cowley and Thomas Sprat, *The Works of Mr. Abraham Cowley: Consisting of Those Which Were Formerly Printed, and Those Which He Design'd for the Press, Now Published Out of the Authors Original Copies* (London: Printed by J. M. for Henry Herringman, 1668), 39–40.

9. Macvey Napier, *Lord Bacon and Sir Walter Raleigh* (Macmillan, 1853), 18.

NINE Demonstration

1. D'Arcy Power, *Masters of Medicine: William Harvey* (T. Fisher Unwin, 1897), 49, 58.

2. Effie Bendann, *Death Customs: An Analytical Study of Burial Rites* (Routledge, 2010), 48–49; James Longrigg, *Greek Rational Medicine: Philosophy and Medicine from Alcmaeon to the Alexandrians* (Routledge, 1993), 184–85.

3. Roy Porter, *The Cambridge Illustrated History of Medicine* (Cambridge University Press, 1988), 75, 157; Lawrence I. Conrad et al., *The Western Medical Tradition: 800 B.C.–1800 A.D.* (Cambridge University Press, 1995), 147.

4. Charles Singer and C. Rabin, *A Prelude to Modern Science* (Cambridge University Press, 1946), xxxiii; Conrad et al., *Western Medical Tradition*, 275–77; Charles Donald O'Malley, *Andreas Vesalius of Brussels, 1514–1564* (University of California Press, 1964), 117.

5. Andreas Vesalius, *On the Fabric of the Human Body. Book VI, The Heart and Associated Organs. Book VII, The Brain: A Translation of De humani corporis fabicra libri septem*, trans. William Frank Richardson and John Burd Carman (Norman, 2009), 83.

6. Power, *Masters of Medicine*, 55–56.

7. Robert C. Olby et al., eds., *Companion to the History of Modern Science* (Routledge, 1990), 569–70; Lois N. Magner, *A History of the Life Sciences*, 3rd ed. (Marcel Dekker, 2002), 83.

8. Magner, *History of the Life Sciences*, 91; Power, *Masters of Medicine*, 149.

9. John G. Simmons, *Doctors and Discoveries: Lives That Created Today's Medicine* (Houghton Mifflin, 2002), 48.

10. Catherine Drinker Bowen, *Francis Bacon: The Temper of a Man* (Little, Brown, 1963), 14.

11. Power, *Masters of Medicine*, 231.

TEN The Death of Aristotle

1. Giorgio de Santillana, *The Crime of Galileo* (University of Chicago Press, 1955), 3.

2. John Joseph Fahie, *Galileo: His Life and Work* (J. Murray, 1903), 27.

3. Stillman Drake, *Galileo at Work: His Scientific Biography* (Dover, 1978), 2, 473.

4. Ibid., 21–22.

5. Galileo Galilei, *Dialogue concerning the Two Chief World Systems, Ptolemaic and Copernican*, trans. Stillman Drake, ed. Stephen Jay Gould (Modern Library, 2001), 125.

6. David Leverington, *Babylon to Voyager and Beyond: A History of Planetary Astronomy* (Cambridge University Press, 2003), 70.

7. William Cecil Dampier and Margaret Dampier, eds., *Cambridge Readings in the Literature of Science; Being Extracts from the Writings of Men of Science to Illustrate the Development of Scientific Thought* (Cambridge University Press, 1928), 15.

8. Maurice A. Finocchiaro, *Defending Copernicus and Galileo: Critical Reasoning in the Two Affairs* (Springer, 2010), xv; Dampier and Dampier, *Cambridge Readings*, 26–27, 30; Leverington, *Babylon to Voyager*, 83.

9. David Deming, *Science and Technology in World History* (McFarland, 2010), 3:165.

10. Galilei, *Dialogue*, 130–31.

11. Galileo Galilei and Maurice A. Finocchiaro, *The Essential Galileo* (Hackett, 2008), 146.

12. Ibid., 147.

13. Galilei, *Dialogue*, xvi, 5, 538.
14. Deming, *Science and Technology*, 177–78.

ELEVEN   Instruments and Helps

1. Thomas Birch, "The Life of the Honourable Robert Boyle," in Robert Boyle, *The Philosophical Works of the Honourable Robert Boyle in Six Volumes* (J. & F. Rivington, 1772), 1:xxiv.
2. Robert Boyle, "A Free Inquiry into the Vulgar Notion of Nature," in *The Philosophical Works of the Honourable Robert Boyle* (W. & J. Innys, 1725), 2:115.
3. Untitled column, *Journal of the Optical Society of America and Review of Scientific Instruments* 6, no. 6 (August 1922): 835–36; Matteo Valleriani, *Galileo Engineer* (Springer, 2010), 56–57.
4. Marie Boas Hall, *Robert Boyle and Seventeenth-Century Chemistry* (Cambridge University Press, 1958), 20.
5. Robert Boyle, *The Sceptical Chymist* (Dover, 2003), 15; Thomas L. Hankins and Robert J. Silverman, *Instruments and the Imagination* (Princeton University Press, 1995), 3.
6. Trevor H. Levere, *Transforming Matter: A History of Chemistry from Alchemy to the Buckyball* (Johns Hopkins University Press, 2001), 14.
7. Birch, "Life of the Honourable Robert Boyle," xxxiv.
8. Hall, *Robert Boyle*, 6; Charles Webster, ed., *The Intellectual Revolution of the Seventeenth Century* (Routledge, 2011), 236–37.
9. Edward Grant, *A Source Book in Medieval Science* (Harvard University Press, 1974), 324, 326.
10. Boyle, *Philosophical Works* (1772), 1:11.
11. Boyle, *Philosophical Works* (1725), 2:510–32; Boyle, *Philosophical Works* (1772), 1:11–12.
12. James Riddick Partington, *A Short History of Chemistry*, 3rd ed. (Dover, 2011), 22–23.
13. Partington, *Short History of Chemistry*, 29, 36; Levere, *Transforming Matter*, 7–8.
14. Robert Boyle, *A Free Enquiry into the Vulgarly Received Notion of Nature*, ed. Edward B. Davis and Michael Hunter (Cambridge University Press, 1996), 114–15.
15. Boyle, *Philosophical Works* (1725), 3:391.
16. Boyle, *Sceptical Chymist*, 17.
17. Michael Hunter, ed., *Robert Boyle Reconsidered* (Cambridge University Press, 2003), 61; Boyle, *Sceptical Chymist*, 3.
18. Hunter, *Robert Boyle Reconsidered*, 72.

19. Robert D. Purrington, *The First Professional Scientist: Robert Hooke and the Royal Society of London* (Birkhäuser, 2009), 34.

20. Margaret 'Espinasse, *Robert Hooke* (University of California Press, 1962), 43–44.

21. Thomas Birch, *The History of the Royal Society of London* (A. Millar, 1757), 3:344–45.

22. David Freedberg, *The Eye of the Lynx: Galileo, His Friends and the Beginnings of Natural History* (University of Chicago Press, 2002), 180.

23. Robert Hooke, "Preface," in *Micrographia* (James Allestry, 1664).

24. Thomas Birch, *The History of the Royal Society of London* (A. Millar, 1756), 1:215ff.

25. Ibid., 262.

26. Hooke, *Micrographia*, Observation 9.

27. Ibid., Preface.

28. Lawrence Principe, "In Retrospect: The Sceptical Chymist," *Nature* 469 (January 6, 2011): 30.

TWELVE Rules of Reasoning

1. Thomas Birch, *The History of the Royal Society of London* (A. Millar, 1756), 2:501.

2. Thomas Birch, *The History of the Royal Society of London* (A. Millar, 1757), 3:1,10.

3. Ibid., 5, 14, 50.

4. Ibid., 269; Charles Hutton, George Shaw, and Richard Pearson, *The Philosophical Transactions of the Royal Society of London* (C. & R. Baldwin, 1809), 2:341; Adrian Johns, "Reading and Experiment in the Early Royal Society," in *Reading, Society, and Politics in Early Modern England*, ed. Kevin Sharpe and Stephen Zwicker (Cambridge University Press, 2003), 260–61.

5. Peter Machamer, ed., *The Cambridge Companion to Galileo* (Cambridge University Press, 1998), 153–54.

6. I. Bernard Cohen, *Revolution in Science* (Harvard University Press, 1985), 163–70.

7. Ron Larson and Bruce Edwards, *Calculus* (Cengage Learning, 2013), 42.

8. James L. Axtell, "Locke, Newton and the Two Cultures," in *John Locke: Problems and Perspectives*, ed. John W. Yolton (Cambridge University Press, 1969), 166–68.

9. Barry Gower, *Scientific Method: A Historical and Philosophical Introduction* (Routledge, 1997), 69.

10. Isaac Newton, *The Principia: Mathematical Principles of Natural Philosophy*, trans. I. Bernard Cohen and Anne Whitman (University of California Press, 1999), 942.

11. Ibid., 943.

THIRTEEN The Genesis of Geology

1. James Oliver Thomson, *History of Ancient Geography* (Biblo & Tannen, 1965), 124ff, 342–43; Duane W. Roller, ed. and trans., *Eratosthenes' Geography* (Princeton University Press, 2010), 161, 263–64.

2. Gian Battista Vai and W. G. E. Caldwell, eds., *The Origins of Geology in Italy* (Geological Society of America, 2006), 158; Gary D. Rosenberg, *The Revolution in Geology from the Renaissance to the Enlightenment* (Geological Society of America, 2010), 143–44.

3. Charles R. Van Hise, "The Problems of Geology," *Journal of Geology* 12, no. 7 (1904): 589–91.

4. G. Brent Dalrymple, *The Age of the Earth* (Stanford University Press, 1991), 21; James Ussher, *Annals of the World* (E. Tyler, 1658), 17.

5. William H. Stiebing, *Ancient Astronauts, Cosmic Collisions and Other Popular Theories* (Prometheus Books, 1984), 5.

6. Rosenberg, *Revolution in Geology*, 144–45.

7. Isaac Newton, *Mathematical Principles of Natural Philosophy*, trans. Andrew Motte (Daniel Adee, 1848), 486.

8. Dalrymple, *Age of the Earth*, 28–29.

9. Benoît de Maillet, *Telliamed, or, The World Explain'd* (W. Pechin, 1797), 194–95; Dalrymple, *Age of the Earth*, 25–29.

10. John R. Gribbin, *The Scientists: A History of Science Told through the Lives of Its Greatest Inventors* (Random House, 2003), 221–23.

11. Georges-Louis Leclerc, Comte de Buffon, *Natural History, General and Particular*, 2nd ed., trans. William Smellie (W. Strahan and T. Cadell, 1785), 1:1.

12. William Whiston, *A New Theory of the Earth, from Its Original, to the Consummation of All Things*, 5th ed. (John Whiston, 1737), 373; David Spadafora, *The Idea of Progress in Eighteenth-Century Britain* (Yale University Press, 1990), 112–13.

13. Buffon, *Natural History*, 1:33–34.

14. Dalrymple, *Age of the Earth*, 29–30.

15. Jacques Roger, *Buffon: A Life in Natural History*, trans. Sarah Lucille Bonnefoi (Cornell University Press, 1997), 187–93.

16. Buffon, *Natural History*, 1:258.

17. Henry Gee, *In Search of Deep Time: Beyond the Fossil Record to a New History of Life* (Cornell University Press, 2001), 2–4.

FOURTEEN The Laws of the New Science

1. Dennis R. Dean, *James Hutton and the History of Geology* (Cornell University Press, 1992), 1–3; John Playfair, *The Works of John Playfair, Esq.* (Archibald Constable, 1822), 4:43–44.

2. Playfair, *Works*, 46.

3. Gian Battista Vai and W. G. E. Caldwell, eds., *The Origins of Geology in Italy* (Geological Society of America, 2006), 59–61: Martin J. S. Rudwick, *Bursting the Limits of Time: The Reconstruction of Geohistory in the Age of Revolution* (University of Chicago Press, 2005), 135.
4. Playfair, *Works*, 12.
5. Ibid., 49–50.
6. Dean, *James Hutton*, 17, 24–25.
7. Charles R. Van Hise, "The Problems of Geology," *Journal of Geology* 12, no. 7 (1904): 614–15.
8. James Hutton, "Theory of the Earth," *Transactions of the Royal Society of Edinburgh* 1 (1788): 301.
9. Ibid., 304.
10. Playfair, *Works*, 63–64.
11. Dean, *James Hutton*, 18, 154.
12. Ibid., 18; J. E. O'Rourke, "A Comparison of James Hutton's Principles of Knowledge and Theory of the Earth," *Isis* 69, no. 1 (March 1978): 19.
13. Jack Repcheck, *The Man Who Found Time: James Hutton and the Discovery of Earth's Antiquity* (Perseus, 2003), 160–61.
14. Martin J. S. Rudwick, *The Meaning of Fossils: Episodes in the History of Palaeontology*, 2nd ed. (University of Chicago Press, 1985), 104; Claudine Cohen, *The Fate of the Mammoth: Fossils, Myth, and History*, trans. William Rodarmor (University of Chicago Press, 2002), 106–8; John Reader, *Missing Links: In Search of Human Origins* (Oxford University Press, 2011), 45.
15. Martin J. S. Rudwick, *Georges Cuvier, Fossil Bones, and Geological Catastrophes: New Translations & Interpretations of the Primary Texts* (University of Chicago Press, 1997), 21; C. L. E. Lewis and S. J. Knell, *The Making of the Geological Society of London* (Geological Society Publishing House, 2009), 77–78.
16. Rudwick, *Georges Cuvier*, 23–24.
17. Ibid., 84–85; Reader, *Missing Links*, 49.
18. Rudwick, *Georges Cuvier*, 168.
19. Ibid., 190; Trevor Palmer, *Perilous Planet Earth: Catastrophes and Catastrophism through the Ages* (Cambridge University Press, 2003), 30.
20. Rudwick, *Georges Cuvier*, 248.

FIFTEEN A Long and Steady History

1. William Buckland, *Vindiciae geologicae: or, The Connexion of Geology with Religion Explained* (Oxford University Press, 1820), 24.
2. Charles Lyell, *Life, Letters, and Journals of Sir Charles Lyell, Bart.*, ed. Katharine M. Lyell (John Murray, 1881), 1:63; J. M. I. Klaver, *Geology and Religious Sentiment: The Effect of Geological Discoveries on English Society and Literature between 1829–1859* (Brill, 1997), 19.

3. Charles Lyell, *Principles of Geology*, ed. James A. Secord (Penguin, 1997), 3, 6; Klaver, *Geology and Religious Sentiment*, 21–22.
4. Lyell, *Life, Letters, and Journals*, 186–87; Klaver, *Geology and Religious Sentiment*, 22, 26.
5. Lyell, *Life, Letters, and Journals*, 234–35.
6. Ibid., 262.
7. Michael Ruse, *The Darwinian Revolution: Science Red in Tooth and Claw*, 2nd ed. (University of Chicago Press, 1999), 17ff; Lyell, *Principles of Geology*, 240–42.
8. Alfred Russel Wallace, *The Wonderful Century: The Age of New Ideas in Science and Invention* (Swan Sonnenschein, 1903), 349.
9. Walter Alvarez, *T. rex and the Crater of Doom* (Princeton University Press, 2008), 51.

SIXTEEN The Unanswered Question

1. Charles Lyell, *Life, Letters, and Journals of Sir Charles Lyell, Bart.*, ed. Katharine M. Lyell (John Murray, 1881), 1:269, 270.
2. Untitled column, *Exeter Flying Post*, October 3, 1844.
3. G. Brent Dalrymple, *The Age of the Earth* (Stanford University Press, 1991), 32–33.
4. Ibid., 69–71; LaVerne Tolley Gurley and William J. Callaway, *Introduction to Radiologic Technology*, 7th ed. (Mosby, 2011), 58–62; Kristin Iverson, *Full Body Burden* (Crown, 2012), 173.
5. Ernest Rutherford, *Radioactive Transformations* (Yale University Press, 1906), 190–91, 194.
6. Don L. Eicher and Arcie Lee McAlester, *The History of the Earth* (Prentice-Hall, 1980), xvi; Cherry Lewis, *The Dating Game: One Man's Search for the Age of the Earth* (Cambridge University Press, 2000), 27.
7. Arthur Holmes, *The Age of the Earth* (London: Harper Brothers, 1913), 17.
8. Lawrence Badash, "The Age-of-the-Earth Debate," *Scientific American* 261, no. 2 (August 1989): 96.
9. Holmes, *Age of the Earth*, 21, 22.
10. Ibid., 11, 164, 166.
11. Ernest Rutherford, James Chadwick, and Charles Drummond Ellis, *Radiations from Radioactive Substances* (Cambridge University Press, 1930), 536.
12. Holmes, *Age of the Earth*, 173.

SEVENTEEN The Return of the Grand Theory

1. Naomi Oreskes, *The Rejection of Continental Drift: Theory and Method in American Earth Science* (Oxford University Press, 1999), 10, 16–17.
2. Edmund A. Mathez and James D. Webster, *The Earth Machine: The Science of a Dynamic Planet* (Columbia University Press, 2004), 87.

3. Oreskes, *Rejection of Continental Drift*, 27, 33.

4. Alfred Wegener, "The Origin of Continents and Oceans," *Living Age*, 8th series, vol. 26 (April/May/June 1922): 657–58; Mathez and Webster, *Earth Machine*, 87.

5. Oreskes, *Rejection of Continental Drift*, 157; H. E. Le Grand, *Drifting Continents and Shifting Theories* (Cambridge University Press, 1988), 65.

6. Alfred Wegener, *The Origin of Continents and Oceans*, trans. John Biram (Dover, 1966), viii.

7. Wegener, "Origin of Continents and Oceans," 658.

8. Wegener, *Origin of Continents and Oceans*, 217.

9. David M. Lawrence, *Upheaval from the Abyss: Ocean Floor Mapping and the Earth Science Revolution* (Rutgers University Press, 2002), 17–18.

10. Mathez and Webster, *Earth Machine*, 90–91.

EIGHTEEN Catastrophe, Redux

1. Victor R. Baker, "The Spokane Flood Debates: Historical Background and Philosophical Perspective," in *History of Geomorphology and Quaternary Geology*, ed. R. H. Grapes, D. R. Oldroyd, and A. Grigelis (Geological Society of London, 2008), 33, 36–37.

2. John Eliot Allen, Marjorie Burns, and Scott Burns, *Cataclysms on the Columbia: The Great Missoula Floods*, 2nd rev. ed. (Ooligan Press, 2009), 56.

3. Baker, "Spokane Flood Debates," 47.

4. Allen, Burns, and Burns, *Cataclysms on the Columbia*, 71–72.

5. Timothy Ferris, "It Came from Outer Space," *New York Times*, May 25, 1997, https://www.nytimes.com/books/97/05/25/reviews/970525.25ferrist.html; Walter Alvarez, *T. rex and the Crater of Doom* (Princeton University Press, 2008), 45, 53.

6. Janine Bourriau, *Understanding Catastrophe: Its Impact on Life on Earth* (Cambridge University Press, 1992), 29.

7. Alvarez, *T. rex*, 42.

8. Luis W. Alvarez et al., "Extraterrestrial Cause for the Cretaceous-Tertiary Extinction," *Science* 208, no. 4448 (June 6, 1980): 1095.

9. Alvarez, *T. rex*, 81–82.

10. Ibid., 12–14.

11. Ibid., ix.

12. Bourriau, *Understanding Catastrophe*, 5.

NINETEEN Biology

1. John Cassell, *Cassell's History of England* (Cassell, Petter, Galpin, 1884), 5:9; Georges Cuvier, "Biographical Memoir of M. de Lamarck," *Edinburgh New Philosophical Journal* 20 (October 1835–April 1836): 8.

2. Martin J. S. Rudwick, *Bursting the Limits of Time: The Reconstruction of Geo-history in the Age of Revolution* (University of Chicago Press, 2005), 390.

3. M. J. S. Hodge, "Lamarck's Science of Living Bodies," *British Journal for the History of Science* 5, no. 4 (December 1971): 325.

4. André Klarsfeld and Frédéric Revah, *The Biology of Death: Origins of Mortality*, trans. Lydia Brady (Cornell University Press, 2004), 7.

5. J. B. Lamarck, *Zoological Philosophy: An Exposition with Regard to the Natural History of Animals*, trans. Hugh Elliot (Macmillan, 1914), 51, 202.

6. Ibid., 2.

7. Ibid., 12, 41, 46.

8. Ibid., 38–39, 60, 175–76; Ernst Mayr, "Lamarck Revisited," *Journal of the History of Biology* 5, no. 1 (Spring 1972): 60–61.

9. Robert J. Richards, *Darwin and the Emergence of Evolutionary Theories of Mind and Behavior* (University of Chicago Press, 1987), 63.

10. A. S. Packard, *Lamarck, the Founder of Evolution: His Life and Work* (Longmans, Green, 1901), 56–58, 70.

TWENTY Natural Selection

1. J. B. Lamarck, *Zoological Philosophy: An Exposition with Regard to the Natural History of Animals*, trans. Hugh Elliot (Macmillan, 1914), 35, 176.

2. Richard A. Richards, *The Species Problem: A Philosophical Analysis* (Cambridge University Press, 2010), 31; Aristotle, *The History of Animals*, trans. Richard Cresswell (Henry G. Bohn, 1862), I.1, sec. 6–8.

3. Monroe W. Strickberger, *Evolution*, 3rd ed. (Jones & Bartlett, 2000), 9.

4. Ibid.

5. Ernst Mayr, *The Growth of Biological Thought: Diversity, Evolution, and Inheritance* (Harvard University Press, 1982), 257–58.

6. Ibid., 394–96; Charles Darwin, *Charles Darwin: His Life Told in an Autobiographical Chapter, and in a Selected Series of His Published Letters*, ed. Francis Darwin (John Murray, 1908), 20.

7. Charles Darwin, *Charles Darwin's Beagle Diary*, ed. R. D. Keynes (Cambridge University Press, 2001), 16.

8. Charles Darwin, *The Origin of Species* (Wordsworth Classics, 1998), 36; Mayr, *Growth of Biological Thought*, 265–66.

9. Frank N. Egerton III, "Darwin's Early Reading of Lamarck," *Isis* 67, no. 3 (September 1976): 453.

10. C. R. Darwin, *Notebook B: [Transmutation of Species (1837–1838)] CUL-DAR121* (transcribed by Kees Rookmaaker), Darwin Online, http://darwin-online.org.uk, accessed May 2014.

11. Darwin, *His Life Told*, 52; Darwin, *Origin of Species*, 186; Charles Darwin, *On Evolution: The Development of the Theory of Natural Selection*, ed. Thomas F. Glick and David Kohn (Hackett, 1996), 83.

12. T. R. Malthus, *Population: The First Essay* (University of Michigan Press, 1959), 4, 6.

13. Darwin, *His Life Told*, 82.

14. Alfred Russel Wallace, *Infinite Tropics: An Alfred Russel Wallace Anthology*, ed. Andrew Berry (Verso, 2002), 51.

15. Darwin, *His Life Told*, 82 ; Mayr, *Growth of Biological Thought*, 423.

16. Mayr, *Growth of Biological Thought*, 423–24.

17. Darwin, *His Life Told*, 42, 46.

18. Untitled column, *Annual Register of World Events: A Review of the Year* 113 (1872): 368.

TWENTY-ONE Inheritance

1. Charles Darwin, *The Origin of Species* (Wordsworth Classics, 1998), 13.

2. Charles Darwin, *The Variation of Animals and Plants under Domestication* (D. Appleton, 1897), 2:371; P. Kyle Stanford, *Exceeding Our Grasp: Science, History, and the Problem of Unconceived Alternatives* (Oxford University Press, 2006), 65.

3. Michael R. Rose, *Darwin's Spectre: Evolutionary Biology in the Modern World* (Princeton University Press, 1998), 33; Peter Atkins, *Galileo's Finger: The Ten Great Ideas of Science* (Oxford University Press, 2004), 45–46.

4. Gregor Mendel, *Experiments in Plant Hybridisation* (Cosimo Classics, 2008), 15, 21ff, 47.

5. Atkins, *Galileo's Finger*, 48–49; Alain F. Corcos and Floyd V. Monaghan, *Gregor Mendel's Experiments on Plant Hybrids: A Guided Study* (Rutgers University Press, 1993), 28–30.

6. J. A. Moore, *Heredity and Development*, 2nd ed. (Oxford University Press, 1972), 29, 45; Atkins, *Galileo's Finger*, 52–55.

7. Rose, *Darwin's Spectre*, 41.

8. Moore, *Heredity and Development*, 74.

TWENTY-TWO Synthesis

1. David Paul Crook, *Darwinism, War and History: The Debate over the Biology of War from the "Origin of Species" to the First World War* (Cambridge University Press, 1994), 1, 15; Raymond Pearl, "Biology and War," *Journal of the Washington Academy of Sciences* 8, no. 11 (June 4, 1918): 355.

2. Alexis de Tocqueville, *Democracy in America* (D. Appleton, 1899), 1:326, 328; Edwin Scott Gaustad and Mark A. Noll, eds., *A Documentary History of Religion in America since 1877*, 3rd ed. (Wm. B. Eerdmans, 2003), 350.

3. Jan Sapp, *Genesis: The Evolution of Biology* (Oxford University Press, 2003), 63; Ernst Mayr and William B. Provine, *The Evolutionary Synthesis: Perspectives on the Unification of Biology* (Harvard University Press, 1998), 3, 8–9.

4. Mayr and Provine, *Evolutionary Synthesis*, 8, 282, 315, 316.

5. T. H. Huxley and Leonard Huxley, *Life and Letters of Thomas Henry Huxley* (D. Appleton, 1900), 1:391.

6. Krishna R. Dronamraju, *If I Am to Be Remembered: The Life and Work of Julian Huxley with Selected Correspondence* (World Scientific, 1993), 5, 9–12, 15.

7. Ibid., 42.

8. Vassiliki Betty Smocovitis, *Unifying Biology: The Evolutionary Synthesis and Evolutionary Biology* (Princeton University Press, 1996), 140.

9. Calendar entry ("Association for the Study of Systematics in Relation to General Biology"), *Nature*, July 24, 1937, 164.

10. John Krige and Dominique Pestre, eds., *Science in the Twentieth Century* (Routledge, 2013), 422.

11. Julian Huxley, *Evolution: The Modern Synthesis*, definitive ed. (MIT Press, 2010), 22, 26–28.

12. Ibid., 3, 6–7.

TWENTY-THREE The Secret of Life

1. James D. Watson, *The Double Helix: A Personal Account of the Discovery of the Structure of DNA* (Scribner, 1993), 197; Daniel D. Chiras, *Human Biology* (Jones & Bartlett, 2013), 357; John C. Kotz, Paul M. Treichel, and John Townsend, *Chemistry and Chemical Reactivity* (Cengage Learning, 2009), 392; Peter Atkins, *Galileo's Finger: The Ten Great Ideas of Science* (Oxford University Press, 2004), 62.

2. Robert Hooke, *Micrographia* (James Allestry, 1664), Observation 18; Robert C. Olby et al., eds., *Companion to the History of Modern Science* (Routledge, 1990), 358–59.

3. Olby et al., *Companion to the History*, 359; Theodor Schwann, *Microscopical Researches into the Accordance in the Structure and Growth of Animals and Plants*, trans. Henry Smith (Sydenham Society, 1847), 242.

4. J. Craig Venter, *Life at the Speed of Light: From the Double Helix to the Dawn of Digital Life* (Viking, 2013), 13; G. P. Talwar and L. M. Srivastava, eds., *Textbook of Biochemistry and Human Biology*, 3rd ed. (Prentice-Hall of India, 2003), xxiv.

5. Joseph Needham, ed., *The Chemistry of Life: Eight Lectures on the History of Biochemistry* (Cambridge University Press, 1970), 17–18.

6. Paul O. P. Ts'o, ed., *Basic Principles in Nucleic Acid Chemistry* (Academic Press, 1974), 1:2; Rudolf Hausmann, *To Grasp the Essence of Life: A History of Molecular Biology* (Kluwer Academic, 2002), 42.

7. Ts'o, *Basic Principles*, 8.

8. David Bainbridge, *The X in Sex: How the X Chromosome Controls Our Lives* (Harvard University Press, 2003), 5.

9. Eric C. R. Reeve, ed., *Encyclopedia of Genetics* (Routledge, 2014), 7.

10. Israel Rosenfield, Edward Ziff, and Borin Van Loon, *DNA: A Graphic Guide to the Molecule That Shook the World* (Columbia University Press, 2011), 3.

11. Isidore Epstein, ed., *Hebrew-English Edition of the Babylonian Talmud: Yebamoth* (Soncino Press, 1984), 48.

12. Hermann Joseph Muller, *The Modern Concept of Nature* (SUNY Press, 1973), 36, 132; Hausmann, *To Grasp the Essence of Life*, 56.

13. William Purves et al., *Life: The Science of Biology*, 7th ed. (Sinauer Associates, 2004), 107, 114, 234; Reeve, *Encyclopedia of Genetics*, 10; Hausmann, *To Grasp the Essence of Life*, 48.

14. Hausmann, *To Grasp the Essence of Life*, 103–4.

15. Ibid., 63–66.

16. Watson, *Double Helix*, 33–35.

17. Ibid., 14–15.

18. Ibid., 20, 50.

19. Ibid., 174, 220.

20. Colin Tudge, *Engineer in the Garden* (Random House, 1993), e-book, chap. 2 subheading "How Does DNA Work?".

21. Francis Crick, *What Mad Pursuit: A Personal View of Scientific Discovery* (Basic Books, 2008), 108–9.

TWENTY-FOUR Biology and Destiny

1. Albert Rosenfeld, "The New Man: What Will He Be Like?" *Life* 59, no. 14 (October 1, 1965): 100.

2. Michael Ruse and Joseph Travis, eds., *Evolution: The First Four Billion Years* (Harvard University Press, 2009), 579–81; Paul S. Agutter and Denys N. Wheatley, *Thinking about Life: The History and Philosophy of Biology and Other Sciences* (Springer, 2008), 194.

3. John H. Gillespie, *Population Genetics: A Concise Guide*, 2nd ed. (Johns Hopkins University Press, 2010), xi.

4. Pierre-Henri Gouyon, Jean-Pierre Henry, and Jacques Arnold, *Gene Avatars: The Neo-Darwinian Theory of Evolution* (Kluwer, 2002), 98.

5. Connie Barlow, ed., *From Gaia to Selfish Genes: Selected Writings in the Life Sciences* (MIT Press, 1992), 156.

6. Gouyon, Henry, and Arnold, *Gene Avatars*, 159–60; Barlow, *From Gaia to Selfish Genes*, 156–57.

7. Steven A. Frank, "The Price Equation, Fisher's Fundamental Theorem, Kin Selection, and Causal Analysis," *Evolution* 51, no. 6 (August 1997): 1713; Kalyanmoy Deb, ed., *Genetic and Evolutionary Computation* (Springer, 2004), 915; Karthik Panchanathan, "George Price, the Price Equation, and Cultural Group Selection," *Evolution and Human Behavior* 32, no. 5 (September 2011): 369, 371.

8. Richard Dawkins, *The Selfish Gene* (Oxford University Press, 1976), 1; Barlow, *From Gaia to Selfish Genes*, 195.

9. Matt Ridley, *The Red Queen: Sex and the Evolution of Human Nature* (Harper Perennial, 2003), 9; Alan Grafen and Mark Ridley, eds., *Richard Dawkins: How a Scientist Changed the Way We Think* (Oxford University Press, 2007), 7.

10. Edward O. Wilson, *Letters to a Young Scientist* (Liveright, 2013), 83–85.

11. Barlow, *From Gaia to Selfish Genes*, 158.

12. Edward O. Wilson, *The Social Conquest of Earth* (W. W. Norton, 2012), 169; Barlow, *From Gaia to Selfish Genes*, 149–50.

13. Edward O. Wilson, *Sociobiology: The New Synthesis* (Harvard University Press, 1975), 3.

14. Ibid., 6.

15. Elizabeth Allen et al., "Against 'Sociobiology,'" *New York Review of Books* 22, no. 18 (November 13, 1975), http://www.nybooks.com/articles/archives/1975/nov/13/against-sociobiology.

16. Edward O. Wilson, *On Human Nature* (Harvard University Press, 2004), xvii.

17. Ibid., 2, 137, 188, 201.

18. Stephen Jay Gould, *The Mismeasure of Man*, rev. and exp. ed. (W. W. Norton, 1996), 20–21.

19. Hans J. Eysenck, *Intelligence: A New Look* (Transaction, 2000), 10.

20. Stephen Jay Gould, *The Richness of Life: The Essential Stephen Jay Gould*, ed. Steven Rose (W. W. Norton, 2007), 446.

21. Ibid., 465–66.

TWENTY-FIVE Relativity

1. Eric Voegelin, *History of Political Ideas*, vol. 6, *Revolution and the New Science* (University of Missouri Press, 1998), 194–95; Nick Hugget, ed., *Space from Zeno to Einstein: Classic Readings with a Contemporary Commentary* (Bradford Books, 1999), 182; George Berkeley, *De motu: Sive de motus principio & natura et de causa communicationis motuum*, trans. A. A. Luce (Jacobi Tonson, 1721), sec. 66.

2. Isaac Newton, *Newton: Philosophical Writings*, ed. Andrew Janiak (University of Cambridge Press, 2004), 100–101.

3. Ioan James, *Remarkable Physicists: From Galileo to Yukawa* (Cambridge University Press, 2004), 69; Charles Coulston Gillispie, *Pierre-Simon Laplace, 1749–1827: A Life in Exact Science* (Princeton University Press, 2000), 273.

4. Edward Harrison, *Cosmology: The Science of the Universe*, 2nd ed. (Cambridge University Press, 2000), 70.

5. Newton, *Philosophical Writings*, 94; Harrison, *Cosmology*, 60–61.

6. Harrison, *Cosmology*, 76; William Huggins, *The Scientific Papers of Sir William Huggins* (W. Wesley and Son, 1909), 221.

7. Eli Maor, *To Infinity and Beyond: A Cultural History of the Infinite* (Princeton University Press, 1991), 131.

8. Ian Stewart, *In Pursuit of the Unknown: 17 Equations That Changed the World* (Basic Books, 2012), 16–17; Jeremy Gray, *Plato's Ghost: The Modernist Transformation of Mathematics* (Princeton University Press, 2008), 48.

9. Michio Kaku, *Hyperspace: A Scientific Odyssey through Parallel Universes, Time Wars, and the 10th Dimension* (Oxford University Press, 1994), 36.

10. Ibid., 338.

11. Peter Galison, Michael Gordin, and David Kaiser, eds., *Science and Society: The History of Modern Physical Science in the Twentieth Century* (Routledge, 2001), 216.

12. Albert Einstein, *Relativity: The Special and General Theory*, trans. Robert W. Lawson (Pi Press, 2005), 19.

13. Ibid., 25, 28.

14. Galison, Gordin, and Kaiser, *Science and Society*, 223; Jay M. Pasachoff and Alex Filippenko, *The Cosmos: Astronomy in the New Millennium*, 4th ed. (Cambridge University Press, 2014), 239–40, 271–72.

15. Pasachoff and Filippenko, *Cosmos*, 240, 274; Albert Einstein and Leopold Infeld, *The Evolution of Physics: The Growth of Ideas from Early Concepts to Relativity and Quanta* (Cambridge University Press, 1938), 310; Maor, *To Infinity and Beyond*, 133.

TWENTY-SIX Damn Quantum Jumps

1. Albert Einstein, quoted in Franco Selleri, *Quantum Paradoxes and Physical Reality: Fundamental Theories of Physics* (Kluwer Academic, 1990), 363.

2. Theodore Arabatzis, *Representing Electrons: A Biographical Approach to Theoretical Entities* (University of Chicago Press, 2006), 56, 61–62; Max Planck, *The Origin and Development of the Quantum Theory*, trans. H. T. Clarke and L. Silberstein (Clarendon Press, 1922), 5.

3. John S. Rigden, *Einstein 1905: The Standard of Greatness* (Harvard University Press, 2005), 68–69.

4. Bernard Fernandez, *Unravelling the Mystery of the Atomic Nucleus: A Sixty-Year Journey, 1896–1956*, trans. Georges Ripka (Springer, 2013), 57–58.

5. Ibid., 58.

6. Ibid., 73; Ernest Rutherford, *The Collected Papers of Lord Rutherford of Nelson* (Interscience, 1963), 2:212.

7. Vern Ostdiek and Donald Bord, *Inquiry into Physics* (Cengage Learning, 2007), 316–17.

8. Bruce Rosenblum and Fred Kuttner, *Quantum Enigma: Physics Encounters Consciousness*, 2nd ed. (Oxford University Press, 2011), 59–60; M. S. Lon-

gair, *Theoretical Concepts in Physics: An Alternative View of Theoretical Reasoning in Physics*, 2nd ed. (Cambridge University Press, 2003), 339.

9. Albert Einstein and Leopold Infeld, *The Evolution of Physics: The Growth of Ideas from Early Concepts to Relativity and Quanta* (Cambridge University Press, 1938), 251.

10. Longair, *Theoretical Concepts in Physics*, 381–83; Einstein and Infeld, *Evolution of Physics*, 267.

11. Planck, *Origin and Development*, 12.

12. Walter J. Moore, *A Life of Erwin Schrödinger* (University of Cambridge Press, 1994), 163; John Gribbin, *Erwin Schrödinger and the Quantum Revolution* (John Wiley & Sons, 2013), 110.

13. T. J. Rice, *Joyce, Chaos, and Complexity* (University of Illinois Press, 1997), 152–53.

14. Einstein and Infeld, *Evolution of Physics*, 273–74.

15. Erwin Schrödinger, "The Present Situation in Quantum Mechanics," trans. John D. Trimmer, *Proceedings of the American Philosophical Society*, November 29, 1935, 328.

16. Gribbin, *Erwin Schrödinger*, 133.

17. Walter J. Moore, *Schrödinger: Life and Thought* (Cambridge University Press, 1992), 404.

TWENTY-SEVEN The Triumph of the Big Bang

1. Robert Bless, *Discovering the Cosmos* (University Science Books, 1996), 527; Jeffrey Crelinsten, *Einstein's Jury: The Race to Test Relativity* (Princeton University Press, 2006), 48, 177–78; John Earman, Michel Janssen, and J. D. Norton, eds., *The Attraction of Gravitation: New Studies in the History of General Relativity* (Center for Einstein Studies, 1993), 161–63.

2. Robert William Smith, *The Expanding Universe: Astronomy's "Great Debate," 1900–1931* (Cambridge University Press, 1982), 112–13; Jay M. Pasachoff and Alex Filippenko, *The Cosmos: Astronomy in the New Millennium*, 4th ed. (Cambridge University Press, 2014), 414; David Levy, ed., *The Scientific American Book of the Cosmos* (St. Martin's Press, 2000), 60.

3. Levy, *Scientific American Book*, 100; Giora Shaviv, *The Synthesis of the Elements: The Astrophysical Quest for Nucleosynthesis and What It Can Tell Us about the Universe* (Springer, 2012), 211–13.

4. Pasachoff and Filippenko, *Cosmos*, 416; William McCrea, "Astronomical Achievements Out of This Galaxy," *New Scientist* 98, no. 1354 (April 21, 1983): 174.

5. Edwin Hubble, *The Realm of the Nebulae* (Yale University Press, 1982), 21; Michio Kaku, *Einstein's Cosmos: How Albert Einstein's Vision Transformed Our Understanding of Space and Time* (W. W. Norton, 2004), 209.

6. Kenneth R. Lang, *Astrophysical Formulae*, vol. 2, *Space, Time, Matter and Cosmology*, 3rd ed. (Springer, 2006), 107.

7. Hubble, *Realm of the Nebulae*, 122.

8. Ibid., 201–2.

9. Kaku, *Einstein's Cosmos*, 123–35; Helge Kragh, *Cosmology and Controversy: The Historical Development of Two Theories of the Universe* (Princeton University Press, 1996), 29–31; David Topper, *How Einstein Created Relativity Out of Physics and Astronomy* (Springer, 2012), 168.

10. Kragh, *Cosmology and Controversy*, 34; Topper, *How Einstein Created Relativity*, 174.

11. Robert M. Wald, *General Relativity* (University of Chicago Press, 1984), 213.

12. Harlow Shapley, ed., *Source Book in Astronomy, 1900–1950* (Harvard University Press, 1960), 363.

13. Milton K. Munitz, ed., *Theories of the Universe: From Babylonian Myth to Modern Science* (Free Press, 1957), 425. The quote is actually from Hoyle's 1950 popularization of his 1948 scientific paper.

14. Simon Mitton, *Fred Hoyle: A Life in Science* (Cambridge University Press, 2011), 128–29.

15. Topper, *How Einstein Created Relativity*, 180.

16. Mitton, *Fred Hoyle*, 116; Ralph A. Alpher and Robert Herman, "'Big-Bang' Cosmology and Cosmic Blackbody Radiation," in *Modern Cosmology in Retrospect*, ed. B. Bertotti et al. (Cambridge University Press, 1990), 147.

17. Charles Seife, *Alpha and Omega: The Search for the Beginning and End of the Universe* (Penguin, 2004), 47; N. Mandolesi and N. Vittorio, eds., *The Cosmic Microwave Background: 25 Years Later* (Kluwer Academic, 1990), 20–24.

18. Bertotti et al., *Modern Cosmology in Retrospect*, 344.

19. Frank Durham and Robert D. Purrington, *Frame of the Universe* (Columbia University Press, 1983), 208.

20. Elizabeth Leane, *Reading Popular Physics: Disciplinary Skirmishes and Textual Strategies* (Ashgate, 2007), 35.

21. Steven Weinberg, *The First Three Minutes: A Modern View of the Origin of the Universe*, 2nd ed. (Basic Books, 1993), 8, 149.

22. Ibid., 153.

23. Leane, *Reading Popular Physics*, 18; Weinberg, *First Three Minutes*, 154–55.

TWENTY-EIGHT The Butterfly Effect

1. Pierre-Simon Laplace, quoted in Leonard Smith, *Chaos: A Very Short Introduction* (Oxford University Press, 2007), 2.

2. H. R. Shaw, *Craters, Cosmos, and Chronicles: A New Theory of Earth* (Stanford University Press, 1995), 387; William E. Doll et al., eds., *Chaos, Com-*

plexity, Curriculum, and Culture: A Conversation (Peter Lang, 2008), 135–37, 154.

3. Doll et al., *Chaos, Complexity*, 154–55.
4. Danette Paul, "Spreading Chaos: The Role of Popularizations in the Diffusion of Scientific Ideas," *Written Communication* 21, no. 1 (January 2004): 37–38; Doll et al., *Chaos, Complexity*, 155.
5. Doll et al., *Chaos, Complexity*, 155.

Works Cited

Agutter, Paul S., and Denys N. Wheatley. *Thinking about Life: The History and Philosophy of Biology and Other Sciences*. Dordrecht, Netherlands: Springer, 2008.

Allen, Elizabeth, Barbara Beckwith, Jon Beckwith, Steven Chorover, David Culver, Margaret Duncan, Steven Gould, et al. "Against 'Sociobiology.'" *New York Review of Books* 22, no. 18 (November 13, 1975), http://www .nybooks .com/articles/archives/1975/nov/13/against-sociobiology.

Allen, John Eliot, Marjorie Burns, and Scott Burns. *Cataclysms on the Columbia: The Great Missoula Floods*, 2nd rev. ed. Portland, OR: Ooligan Press, 2009.

Alpher, Ralph A., and Robert Herman, "Early Work on 'Big-Bang' Cosmology and Cosmic Blackbody Radiation." In *Modern Cosmology in Retrospect*. Edited by B. Bertotti, R. Balbinot, S. Bergia, and A. Messina. Cambridge: Cambridge University Press, 1990.

Alvarez, Luis W., Walter Alvarez, Frank Asaro, and Helen V. Michel. "Extra-terrestrial Cause for the Cretaceous-Tertiary Extinction." *Science* 208, no. 4448 (June 6, 1980): 1095–1108.

Alvarez, Walter. *T. rex and the Crater of Doom*. Princeton, NJ: Princeton University Press, 2008.

Annual Register of World Events: A Review of the Year 113 (1872).

Arabatzis, Theodore. *Representing Electrons: A Biographical Approach to Theoretical Entities*. Chicago: University of Chicago Press, 2006.

Archimedes. "The Sand-Reckoner." In *The Works of Archimedes*. Translated by Thomas. L. Heath. Cambridge: Cambridge University Press, 1897.

Aristotle. *The History of Animals*. Translated by Richard Cresswell. London: Henry G. Bohn, 1862.

———. *The Metaphysics*. Translated by William David Ross. In *The Works of Aristotle*, vol. 3. Franklin Center, PA: Franklin Library, 1982.

———. *Politics*. Translated by Ernest Barker. Oxford World's Classics. Oxford: Oxford University Press, 1988.

Atkins, Peter. *Galileo's Finger: The Ten Great Ideas of Science.* Oxford: Oxford University Press, 2004.

Axtell, James L. "Locke, Newton and the Two Cultures." In *John Locke: Problems and Perspectives.* Edited by John W. Yolton. Cambridge: Cambridge University Press, 1969.

Bacon, Francis. *The Philosophical Works of Francis Bacon in Five Volumes*, vol. 4. Edited by James Spedding. London: Longman, 1861.

Badash, Lawrence. "The Age-of-the-Earth Debate." *Scientific American* 261, no. 2 (August 1989).

Bainbridge, David. *The X in Sex: How the X Chromosome Controls Our Lives.* Cambridge, MA: Harvard University Press, 2003.

Baker, Victor R. "The Spokane Flood Debates: Historical Background and Philosophical Perspective." In *History of Geomorphology and Quaternary Geology.* Edited by R. H. Grapes, D. R. Oldroyd, and A. Grigelis. London: Geological Society of London, 2008.

Barlow, Connie, ed. *From Gaia to Selfish Genes: Selected Writings in the Life Sciences.* Cambridge, MA: MIT Press, 1992.

Barnes, Jonathan, ed. *The Cambridge Companion to Aristotle.* Cambridge: Cambridge University Press, 1995.

———. *Early Greek Philosophy*, rev. ed. New York: Penguin, 2002.

Bauer, Susan Wise. *The History of the Renaissance World.* New York: W. W. Norton, 2013.

Bendann, Effie. *Death Customs: An Analytical Study of Burial Rites.* New York: Routledge, 2010.

Berkeley, George. *De motu: Sive de motus principio & natura et de causa communicationis motuum.* Translated by A. A. Luce. London: Jacob Tonson, 1721.

Birch, Thomas. *The History of the Royal Society of London*, vol. 1. London: A. Millar, 1756.

———. *The History of the Royal Society of London*, vol. 3. London: A. Millar, 1757.

Bless, Robert. *Discovering the Cosmos.* Sausalito, CA: University Science Books, 1996.

Bourriau, Janine. *Understanding Catastrophe: Its Impact on Life on Earth.* Cambridge: Cambridge University Press, 1992.

Bowen, Catherine Drinker. *Francis Bacon: The Temper of a Man.* Boston: Little, Brown, 1963.

Boyle, Robert. *A Free Enquiry into the Vulgarly Received Notion of Nature.* Edited by Edward B. Davis and Michael Hunter. Cambridge: Cambridge University Press, 1996.

———. *The Philosophical Works of the Honourable Robert Boyle*, vols. 2 and 3. London: W. & J. Innys, 1725.

————. *The Philosophical Works of the Honourable Robert Boyle in Six Volumes,* vol. 1. London: J. & F. Rivington, 1772.

————. *The Sceptical Chymist.* New York: Dover, 2003.

Buckland, William. *Vindiciae geologicae: or, The Connexion of Geology with Religion Explained.* Oxford: Oxford University Press, 1820.

Buffon, Georges-Louis Leclerc, Comte de. *Natural History, General and Particular,* 2nd ed., vol. 1. Translated by William Smellie. London: W. Strahan and T. Cadell, 1785.

Cassell, John. *Cassell's History of England,* vol. 5. London: Cassell, Petter, Galpin, 1884.

Chiras, Daniel D. *Human Biology.* Sudbury, MA: Jones & Bartlett, 2013.

Cohen, Claudine. *The Fate of the Mammoth: Fossils, Myth, and History.* Translated by William Rodarmor. Chicago: University of Chicago Press, 2002.

Cohen, H. Floris. *How Modern Science Came into the World: Four Civilizations, One 17th-Century Breakthrough.* Amsterdam: Amsterdam University Press, 2010.

Cohen, I. Bernard. *Revolution in Science.* Cambridge, MA: Harvard University Press, 1985.

Cohen, S. Marc, Patricia Curd, and C. D. C. Reeve, eds. *Readings in Ancient Greek Philosophy: From Thales to Aristotle,* 4th ed. Indianapolis, IN: Hackett, 2011.

Conrad, Lawrence I., Michael Neve, Vivian Nutton, Roy Porter, and Andrew Wear. *The Western Medical Tradition: 800 B.C.–1800 A.D.* Cambridge: Cambridge University Press, 1995.

Copernicus, Nicolaus. *On the Revolutions of the Heavenly Spheres.* Translated by Charles Glenn Wallis. Amherst, NY: Prometheus Books, 1995.

————. *Three Copernican Treatises.* Translated by Edward Rosen. New York: Dover, 1959.

Corcos, Alain F., and Floyd V. Monaghan. *Gregor Mendel's Experiments on Plant Hybrids: A Guided Study.* New Brunswick, NJ: Rutgers University Press, 1993.

Coyne, George, and Michael Heller. *A Comprehensible Universe.* Dordrecht, Netherlands: Springer, 2008.

Crelinsten, Jeffrey. *Einstein's Jury: The Race to Test Relativity.* Princeton, NJ: Princeton University Press, 2006.

Crick, Francis. *What Mad Pursuit: A Personal View of Scientific Discovery.* New York: Basic Books, 2008.

Crook, David Paul. *Darwinism, War and History: The Debate over the Biology of War from the "Origin of Species" to the First World War.* Cambridge: Cambridge University Press, 1994.

Cuvier, Georges. "Biographical Memoir of M. de Lamarck." *Edinburgh New Philosophical Journal* 20 (October 1835–April 1836): 1–21.

Dalrymple, G. Brent. *The Age of the Earth*. Stanford, CA: Stanford University Press, 1991.

Dampier, William Cecil, and Margaret Dampier, eds. *Cambridge Readings in the Literature of Science; Being Extracts from the Writings of Men of Science to Illustrate the Development of Scientific Thought*. Cambridge: Cambridge University Press, 1928.

Darwin, Charles. *Charles Darwin: His Life Told in an Autobiographical Chapter, and in a Selected Series of His Published Letters*. Edited by Francis Darwin. London: John Murray, 1908.

————. *Charles Darwin's Beagle Diary*. Edited by R. D. Keynes. Cambridge: Cambridge University Press, 2001.

————. *Notebook B: [Transmutation of Species (1837–1838)] CUL-DAR121*. Transcribed by Kees Rookmaaker. Darwin Online, http://darwin-online.org .uk, accessed May 2014.

————. *On Evolution: The Development of the Theory of Natural Selection*. Edited by Thomas F. Glick and David Kohn. Indianapolis, IN: Hackett, 1996.

————. *The Origin of Species*, chap. 2. Hertfordshire, England: Wordsworth Classics, 1998.

————. *The Variation of Animals and Plants under Domestication*, vol. 2. London: D. Appleton, 1897.

Dawkins, Richard. *The Selfish Gene*. Oxford: Oxford University Press, 1976.

Dean, Dennis R. *James Hutton and the History of Geology*. Ithaca, NY: Cornell University Press, 1992.

Deb, Kalyanmoy, ed. *Genetic and Evolutionary Computation*. Dordrecht, Netherlands: Springer, 2004.

Deming, David. *Science and Technology in World History*, vol. 3. Jefferson, NC: McFarland, 2010.

Devlin, Keith. *The Language of Mathematics: Making the Invisible Visible*. New York: W. H. Freeman, 2000.

Doll, William E., M. Jayne Fleener, Donna Trueit, and John St. Julien, eds. *Chaos, Complexity, Curriculum, and Culture: A Conversation*. New York: Peter Lang, 2005.

Drake, Stillman. *Galileo at Work: His Scientific Biography*. New York: Dover, 1978.

Dronamraju, Krishna R. *If I Am to Be Remembered: The Life and Work of Julian Huxley with Selected Correspondence*. River Edge, NJ: World Scientific, 1993.

Durham, Frank, and Robert D. Purrington. *Frame of the Universe*. New York: Columbia University Press, 1983.

Earman, John, Michel Janssen, and J. D. Norton, eds. *The Attraction of Gravitation: New Studies in the History of General Relativity*. Boston: Center for Einstein Studies, 1993.

Egerton, Frank N., III. "Darwin's Early Reading of Lamarck." *Isis* 67, no. 3 (September 1976): 452–56.

Eicher, Don L., and Arcie Lee McAlester. *The History of the Earth.* Englewood Cliffs, NJ: Prentice-Hall, 1980.

Einstein, Albert. *Relativity: The Special and General Theory.* Translated by Robert W. Lawson. New York: Pi Press, 2005.

Einstein, Albert, and Leopold Infeld. *The Evolution of Physics: The Growth of Ideas from Early Concepts to Relativity and Quanta.* Cambridge: Cambridge University Press, 1938.

Epicurus. *Letters and Sayings of Epicurus.* Translated by Odysseus Makridis. New York: Barnes & Noble, 2005.

Epstein, Isidore, ed. *Hebrew-English Edition of the Babylonian Talmud: Yebamoth.* London: Soncino Press, 1984.

'Espinasse, Margaret. *Robert Hooke.* Berkeley: University of California Press, 1962.

Euclid. *The Thirteen Books of the Elements*, 2nd ed. Translated by Thomas L. Heath. Cambridge: Cambridge University Press, 1908.

Eysenck, Hans. J. *Intelligence: A New Look.* New Brunswick, NJ: Transaction, 2000.

Fahie, John Joseph. *Galileo: His Life and Work.* London: J. Murray, 1903.

Ferguson, Kitty. *Measuring the Universe: Our Historic Quest to Chart the Horizons of Space and Time.* New York: Walker, 1999.

Ferris, Timothy. "It Came from Outer Space." *New York Times,* May 25, 1997. https://www.nytimes.com/books/97/05/25/reviews/970525.25ferrist .html.

Finocchiaro, Maurice A. *Defending Copernicus and Galileo: Critical Reasoning in the Two Affairs.* Dordrecht, Netherlands: Springer, 2010.

Frank, Steven A. "The Price Equation, Fisher's Fundamental Theorem, Kin Selection, and Causal Analysis." *Evolution* 51, no. 6 (August 1997): 1712–29.

Freedberg, David. *The Eye of the Lynx: Galileo, His Friends, and the Beginnings of Natural History.* Chicago: University of Chicago Press, 2002.

Galilei, Galileo. *Dialogue concerning the Two Chief World Systems, Ptolemaic and Copernican.* Translated by Stillman Drake. Edited by Stephen Jay Gould. New York: Modern Library, 2001.

Galilei, Galileo, and Maurice A. Finocchiaro. *The Essential Galileo.* Indianapolis, IN: Hackett, 2008.

Galison, Peter, Michael Gordin, and David Kaiser, eds. *Science and Society: The History of Modern Physical Science in the Twentieth Century.* New York: Routledge, 2001.

Gaustad, Edwin Scott, and Mark A. Noll, eds. *A Documentary History of Religion in America since 1877*, 3rd ed. Grand Rapids, MI: Wm. B. Eerdmans, 2003.

Gee, Henry. *In Search of Deep Time: Beyond the Fossil Record to a New History of Life*. Ithaca, NY: Cornell University Press, 2001.

Gilder, Joshua, and Anne-Lee Gilder. *Heavenly Intrigue: Johannes Kepler, Tycho Brahe, and the Murder behind One of History's Greatest Scientific Discoveries*. New York: Random House, 2004.

Gillespie, John H. *Population Genetics: A Concise Guide*, 2nd ed. Baltimore: Johns Hopkins University Press, 2010.

Gillispie, Charles Coulston. *Pierre-Simon Laplace, 1749–1827: A Life in Exact Science*. Princeton, NJ: Princeton University Press, 2000.

Gottlieb, Anthony. *The Dream of Reason: A History of Philosophy from the Greeks to the Renaissance*. New York: W. W. Norton, 2000.

Gould, Stephen Jay. *The Mismeasure of Man*, rev. and exp. ed. New York: W. W. Norton, 1996.

———. *The Richness of Life: The Essential Stephen Jay Gould*. Edited by Steven Rose. New York: W. W. Norton, 2007.

Gouyon, Pierre-Henri, Jean-Pierre Henry, and Jacques Arnold. *Gene Avatars: The Neo-Darwinian Theory of Evolution*. Dordrecht, Netherlands: Kluwer, 2002.

Gower, Barry. *Scientific Method: A Historical and Philosophical Introduction*. New York: Routledge, 1997.

Grafen, Alan, and Mark Ridley, eds. *Richard Dawkins: How a Scientist Changed the Way We Think*. Oxford: Oxford University Press, 2007.

Grant, Edward. *A Source Book in Medieval Science*. Cambridge, MA: Harvard University Press, 1974.

Grattan-Guinness, I., ed. *Companion Encyclopedia of the History & Philosophy of the Mathematical Sciences*. New York: Routledge, 1994.

Gray, Jeremy. *Plato's Ghost: The Modernist Transformation of Mathematics*. Princeton, NJ: Princeton University Press, 2008.

Gribbin, John. *Erwin Schrödinger and the Quantum Revolution*. New York: John Wiley & Sons, 2013.

———. *The Scientists: A History of Science Told through the Lives of Its Greatest Inventors*. New York: Random House, 2003.

Guicciardini, Niccolo. *Isaac Newton on Mathematical Certainty and Method*. Cambridge, MA: MIT Press, 2009.

Gurley, LaVerne Tolley, and William J. Callaway. *Introduction to Radiologic Technology*, 7th ed. St. Louis: Mosby, 2011.

Guthrie, Kenneth S., and David R. Fideler. *The Pythagorean Sourcebook and Library: An Anthology of Ancient Writings Which Relate to Pythagoras and Pythagorean Philosophy*. Newburyport, MA: Phanes Press, 1987.

Hall, Marie Boas. *Robert Boyle and Seventeenth-Century Chemistry*. Cambridge: Cambridge University Press, 1958.

Hankins, Thomas L., and Robert J. Silverman. *Instruments and the Imagination.* Princeton, NJ: Princeton University Press, 1995.

Harrison, Edward. *Cosmology: The Science of the Universe,* 2nd ed. Cambridge: Cambridge University Press, 2000.

Hausmann, Rudolf. *To Grasp the Essence of Life: A History of Molecular Biology.* Dordrecht, Netherlands: Kluwer Academic, 2002.

Hetherington, Norriss S. *Cosmology: Historical, Literary, Philosophical, Religious, and Scientific Perspectives.* New York: CRC Press, 1993.

Hippocrates. *The Corpus: Hippocratic Writings.* Translated by Conrad Fischer. New York: Kaplan, 2008.

———. *On Ancient Medicine.* Translated by Mark J. Schiefsky. Leiden, Netherlands: Brill, 2005.

Hirshfeld, Alan W. *Parallax: The Race to Measure the Cosmos.* Boston: Birkhäuser, 2000.

Hodge, M. J. S. "Lamarck's Science of Living Bodies." *British Journal for the History of Science* 5, no. 4 (December 1971): 323–52.

Hooke, Robert. *Micrographia.* London: James Allestry, 1664.

Hubble, Edwin. *The Realm of the Nebulae.* New Haven, CT: Yale University Press, 1982.

Huffman, Carl. *Archytas of Tarentum: Pythagorean, Philosopher and Mathematician King.* Cambridge: Cambridge University Press, 2005.

Hugget, Nick, ed. *Space from Zeno to Einstein: Classic Readings with a Contemporary Commentary.* Boston: Bradford Books, 1999.

Huggins, William. *The Scientific Papers of Sir William Huggins.* London: W. Wesley and Son, 1909.

Hunter, Michael. *Establishing the New Science: The Experience of the Early Royal Society.* Suffolk, England: Boydell Press, 1989.

———, ed. *Robert Boyle Reconsidered.* Cambridge: Cambridge University Press, 2003.

Hutton, Charles, George Shaw, and Richard Pearson. *The Philosophical Transactions of the Royal Society of London,* vol. 2. London: C. & R. Baldwin, 1809.

Hutton, James. "Theory of the Earth." *Transactions of the Royal Society of Edinburgh* 1 (1788): 209–304.

Huxley, Julian. *Evolution: The Modern Synthesis,* definitive ed. Cambridge, MA: MIT Press, 2010.

Huxley, T. H., and Leonard Huxley. *Life and Letters of Thomas Henry Huxley,* vol. 1. London: D. Appleton, 1900.

Iverson, Kristin. *Full Body Burden.* New York: Crown, 2012.

Jaeger, Mary. *Archimedes and the Roman Imagination.* Ann Arbor: University of Michigan Press, 2008.

James, Ioan. *Remarkable Physicists: From Galileo to Yukawa*. Cambridge: Cambridge University Press, 2004.

Johns, Adrian. "Reading and Experiment in the Early Royal Society." In *Reading, Society, and Politics in Early Modern England*. Edited by Kevin Sharpe and Stephen Zwicker. Cambridge: Cambridge University Press, 2003.

Journal of the Optical Society of America and Review of Scientific Instruments 6, no. 6 (August 1922).

Jowett, Benjamin. *The Dialogues of Plato in Four Volumes*, vol. 2. New York: Charles Scribner's Sons, 1892.

———. *The Dialogues of Plato in Four Volumes*, vol. 4. New York: Hearst's International Library Co., 1914.

Kaku, Michio. *Einstein's Cosmos: How Albert Einstein's Vision Transformed Our Understanding of Space and Time*. New York: W. W. Norton, 2004.

———. *Hyperspace: A Scientific Odyssey through Parallel Universes, Time Wars, and the 10th Dimension*. Oxford: Oxford University Press, 1994.

Kendig, Keith. *Sink or Float: Thought Problems in Math and Physics*. Washington, DC: Mathematical Association of Virginia, 2008.

Kirkpatrick, Larry, and Gregory Francis. *Physics: A World View*. Belmont, CA: Thomson, 2007.

Klarsfeld, André, and Frédéric Revah. *The Biology of Death: Origins of Mortality*. Translated by Lydia Brady. Ithaca, NY: Cornell University Press, 2004.

Klaver, J. M. I. *Geology and Religious Sentiment: The Effect of Geological Discoveries on English Society and Literature between 1829–1859*. Leiden, Netherlands: Brill, 1997.

Kotz, John C., Paul M. Treichel, and John Townsend. *Chemistry and Chemical Reactivity*. Independence, KY: Cengage Learning, 2009.

Kragh, Helge. *Cosmology and Controversy: The Historical Development of Two Theories of the Universe*. Princeton, NJ: Princeton University Press, 1996.

Krige, John, and Dominique Pestre, eds. *Science in the Twentieth Century*. New York: Routledge, 2013.

Kuhn, Thomas S. *The Copernican Revolution: Planetary Astronomy in the Development of Western Thought*. Cambridge, MA: Harvard University Press, 1957.

Lamarck, J. B. *Zoological Philosophy: An Exposition with Regard to the Natural History of Animals*. Translated by Hugh Elliot. London: Macmillan, 1914.

Lang, Kenneth R. *Astrophysical Formulae*. Vol. 2, *Space, Time, Matter and Cosmology*, 3rd ed. Dordrecht, Netherlands: Springer, 2006.

Larson, Ron, and Bruce Edwards. *Calculus*. Independence, KY: Cengage Learning, 2013.

Lawrence, David M. *Upheaval from the Abyss: Ocean Floor Mapping and the Earth Science Revolution*. New Brunswick, NJ: Rutgers University Press, 2002.

Leane, Elizabeth. *Reading Popular Physics: Disciplinary Skirmishes and Textual Strategies*. Surrey, England: Ashgate, 2007.

Le Grand, H. E. *Drifitng Continents and Shifting Theories*. Cambridge: Cambridge University Press, 1988.

Levere, Trevor H. *Transforming Matter: A History of Chemistry from Alchemy to the Buckyball*. Baltimore: Johns Hopkins University Press, 2001.

Leverington, David. *Babylon to Voyager and Beyond: A History of Planetary Astronomy*. Cambridge: Cambridge University Press, 2003.

Levy, David, ed. *The Scientific American Book of the Cosmos*. New York: St. Martin's Press, 2000.

Lewis, C. L. E., and S. J. Knell. *The Making of the Geological Society of London*. London: Geological Society Publishing House, 2009.

Lewis, Cherry. *The Dating Game: One Man's Search for the Age of the Earth*. Cambridge: Cambridge University Press, 2000.

Lindberg, David C. *The Beginnings of Western Science*, 2nd ed. Chicago: University of Chicago Press, 2007.

Linton, C. M. *From Eudoxus to Einstein: A History of Mathematical Astronomy*. Cambridge: Cambridge University Press, 2008.

Longair, M. S. *Theoretical Concepts in Physics: An Alternative View of Theoretical Reasoning in Physics*, 2nd ed. Cambridge: Cambridge University Press, 2003.

Longrigg, James. *Greek Rational Medicine: Philosophy and Medicine from Alcmaeon to the Alexandrians*. New York: Routledge, 1993.

Lucretius. *On the Nature of the Universe*, rev. sub. ed. Translated by Ronald E. Latham. New York: Penguin Classics, 1994.

———. *On the Nature of the Universe*. Translated by Ronald Melville. Oxford: Oxford University Press, 1997.

Lucretius Carus, Titus. *On the Nature of Things*. Translated by John Selby Watson. London: Henry G. Bohn, 1851.

Lyell, Charles. *Life, Letters, and Journals of Sir Charles Lyell, Bart.*, vol. 1. Edited by Katharine M. Lyell. London: John Murray, 1881.

Machamer, Peter, ed. *The Cambridge Companion to Galileo*. Cambridge: Cambridge University Press, 1998.

Magner, Lois N. *A History of the Life Sciences*, 3rd ed. New York: Marcel Dekker, 2002.

Maillet, Benoît de. *Telliamed, or, The World Explain'd*. Baltimore: W. Pechin, 1797.

Malthus, T. R. *Population: The First Essay*. Ann Arbor: University of Michigan Press, 1959.

Mandolesi, N., and N. Vittorio, eds. *The Cosmic Microwave Background: 25 Years Later*. Dordrecht, Netherlands: Kluwer Academic, 1990.

Mankiewicz, Richard. *The Story of Mathematics*. Princeton, NJ: Princeton University Press, 2000.

Maor, Eli. *To Infinity and Beyond: A Cultural History of the Infinite*. Princeton, NJ: Princeton University Press, 1991.

Mathez, Edmund A., and James D. Webster. *The Earth Machine: The Science of a Dynamic Planet*. New York: Columbia University Press, 2004.

Mayr, Ernst. *The Growth of Biological Thought: Diversity, Evolution, and Inheritance*. Cambridge, MA: Harvard University Press, 1982.

———. "Lamarck Revisited." *Journal of the History of Biology* 5, no. 1 (Spring 1972): 55–94.

Mayr, Ernst, and William B. Provine. *The Evolutionary Synthesis: Perspectives on the Unification of Biology*. Cambridge, MA: Harvard University Press, 1998.

McCrea, William. "Astronomical Achievements Out of This Galaxy." *New Scientist* 98, no. 1354 (April 21, 1983): 174.

McElhinny, Michael W., and Phillip L. McFadden. *Paleomagnetism: Continents and Oceans*. New York: Academic Press, 2000.

Mendel, Gregor. *Experiments in Plant Hybridisation*. New York: Cosimo Classics, 2008.

Mensch, Jennifer. *Kant's Organicism: Epigenesis and the Development of Critical Philosophy*. Chicago: University of Chicago Press, 2013.

Mitton, Simon. *Fred Hoyle: A Life in Science*. Cambridge: Cambridge University Press, 2011.

Moore, J. A. *Heredity and Development*, 2nd ed. Oxford: Oxford University Press, 1972.

Moore, Walter J. *A Life of Erwin Schrödinger*. Cambridge: University of Cambridge Press, 1994.

———. *Schrödinger: Life and Thought*. Cambridge: Cambridge University Press, 1992.

Muller, Hermann Joseph. *The Modern Concept of Nature*. Albany, NY: SUNY Press, 1973.

Munitz, Milton K., ed. *Theories of the Universe: From Babylonian Myth to Modern Science*. New York: Free Press, 1957.

Naddaf, Gerard. *The Greek Concept of Nature*. Albany, NY: SUNY Press, 1995.

Napier, Macvey. *Lord Bacon and Sir Walter Raleigh*. New York: Macmillan, 1853.

Needham, Joseph, ed. *The Chemistry of Life: Eight Lectures on the History of Biochemistry*. Cambridge: Cambridge University Press, 1970.

Newton, Isaac. *Mathematical Principles of Natural Philosophy*. Translated by Andrew Motte. London: Daniel Adee, 1848.

———. *Newton: Philosophical Writings*. Edited by Andrew Janiak. Cambridge: University of Cambridge Press, 2004.

———. *The Principia: Mathematical Principles of Natural Philosophy*. Translated by

I. Bernard Cohen and Anne Whitman. Berkeley: University of California Press, 1999.

Olby, Robert C., G. N. Cantor, J. R. R. Christie, and M. J. S. Hodge, eds. *Companion to the History of Modern Science*. New York: Routledge, 1990.

O'Malley, Charles Donald. *Andreas Vesalius of Brussels, 1514–1564*. Berkeley: University of California Press, 1964.

Oreskes, Naomi. *The Rejection of Continental Drift: Theory and Method in American Earth Science*. Oxford: Oxford University Press, 1999.

O'Rouke, J. E. "A Comparison of James Hutton's Principles of Knowledge and Theory of the Earth." *Isis* 69, no. 1 (March 1978): 4–20.

Osler, Margaret J. *Reconfiguring the World: Nature, God, and Human Understanding from the Middle Ages to Early Modern Europe*. Baltimore: Johns Hopkins University Press, 2010.

Ostdiek, Vern, and Donald Bord. *Inquiry into Physics*. Independence, KY: Cengage Learning, 2007.

Packard, A. S. *Lamarck, the Founder of Evolution: His Life and Work*. London: Longmans, Green, 1901.

Palmer, Trevor. *Perilous Planet Earth: Catastrophes and Catastrophism through the Ages*. Cambridge: Cambridge University Press, 2003.

Panchanathan, Karthik. "George Price, the Price Equation, and Cultural Group Selection." *Evolution and Human Behavior* 32, no. 5 (September 2011): 368–71.

Parker, Robert. *On Greek Religion*. Ithaca, NY: Cornell University Press, 2011.

Partington, James Riddick. *A Short History of Chemistry*, 3rd ed. New York: Dover, 2011.

Pasachoff, Jay M., and Alex Filippenko. *The Cosmos: Astronomy in the New Millennium*, 4th ed. Cambridge: Cambridge University Press, 2014.

Paul, Danette. "Spreading Chaos: The Role of Popularizations in the Diffusion of Scientific Ideas." *Written Communication* 21, no. 1 (January 2004): 32–68.

Pausanius. *Pausanias's Description of Greece*, vol. 3. Translated by J. G. Frazer. London: Macmillan, 1898.

Pearl, Raymond. "Biology and War." *Journal of the Washington Academy of Sciences* 8, no. 11 (June 4, 1918): 341–60.

Placher, William C. *A History of Christian Theology: An Introduction*. Louisville, KY: John Knox Press, 1983.

Planck, Max. *The Origin and Development of the Quantum Theory*. Translated by H. T. Clarke and L. Silberstein. Oxford: Clarendon Press, 1922.

Plato, *Plato's Timaeus: Translation, Glossary, Appendices, and Introductory Essay*. Translated by Peter Kalkavage. Newburyport, MA: Focus, 2001.

———. *Protagoras*. Translated by Benjamin Jowett. Rockville, MD: Serenity, 2009.

————. *The Republic: The Complete and Unabridged Jowett Translation*. New York: Vintage, 1991.

Playfair, John. *The Works of John Playfair, Esq.*, vol. 4. London: Archibald Constable, 1822.

Porter, Roy. *The Cambridge Illustrated History of Medicine*. Cambridge: Cambridge University Press, 1988.

Power, D'Arcy. *Masters of Medicine: William Harvey*. London: T. Fisher Unwin, 1897.

Principe, Lawrence. "In Retrospect: The Sceptical Chymist." *Nature* 469 (January 6, 2011): 30–31.

Prioreschi, Plinio. *A History of Medicine*. Vol. 1, *Primitive and Ancient Medicine*, 2nd ed. Omaha, NE: Horatius Press, 1996.

Purrington, Robert D. *The First Professional Scientist: Robert Hooke and the Royal Society of London*. Basel, Switzerland: Birkhäuser, 2009.

Purves, William, David Sadava, Gordon H. Orians, and H. Craig Heller. *Life: The Science of Biology*, 7th ed. Sunderland, MA: Sinauer Associates, 2004.

Reader, John. *Missing Links: In Search of Human Origins*. Oxford: Oxford University Press, 2011.

Reeve, Eric C. R., ed. *Encyclopedia of Genetics*. New York: Routledge, 2014.

Repcheck, Jack. *Copernicus' Secret: How the Scientific Revolution Began*. New York: Simon & Schuster, 2007.

————. *The Man Who Found Time: James Hutton and the Discovery of Earth's Antiquity*. Cambridge, MA: Perseus, 2003.

Rice, T. J. *Joyce, Chaos, and Complexity*. Urbana: University of Illinois Press, 1997.

Richards, Richard A. *The Species Problem: A Philosophical Analysis*. Cambridge: Cambridge University Press, 2010.

Richards, Robert J. *Darwin and the Emergence of Evolutionary Theories of Mind and Behavior*. Chicago: University of Chicago Press, 1987.

Ridley, Matt. *The Red Queen: Sex and the Evolution of Human Nature*. New York: Harper Perennial, 2003.

Rigden, John S. *Einstein 1905: The Standard of Greatness*. Cambridge, MA: Harvard University Press, 2005.

Rochberg, Francesca. *The Heavenly Writing: Divination, Horoscopy, and Astronomy in Mesopotamian Culture*. Cambridge: Cambridge University Press, 2004.

Roger, Jacques. *Buffon: A Life in Natural History*. Translated by Sarah Lucille Bonnefoi. Ithaca, NY: Cornell University Press, 1997.

Roller, Duane W., ed. and trans. *Eratosthenes' Geography*. Princeton, NJ: Princeton University Press, 2010.

Rose, Michael R. *Darwin's Spectre: Evolutionary Biology in the Modern World.* Princeton, NJ: Princeton University Press, 1998.

Rosenberg, Gary D., ed. *The Revolution in Geology from the Renaissance to the Enlightenment.* Boulder, CO: Geological Society of America, 2010.

Rosenblum, Bruce, and Fred Kuttner. *Quantum Enigma: Physics Encounters Consciousness,* 2nd ed. Oxford: Oxford University Press, 2011.

Rosenfeld, Albert. "The New Man: What Will He Be Like?" *Life* 59, no. 14 (October 1, 1965): 94–111.

Rosenfield, Israel, Edward Ziff, and Borin Van Loon. *DNA: A Graphic Guide to the Molecule That Shook the World.* New York: Columbia University Press, 2011.

Rudwick, Martin J. S. *Bursting the Limits of Time: The Reconstruction of Geohistory in the Age of Revolution.* Chicago: University of Chicago Press, 2005.

———. *Georges Cuvier, Fossil Bones, and Geological Catastrophes: New Translations & Interpretations of the Primary Texts.* Chicago: University of Chicago Press, 1997.

———. *The Meaning of Fossils: Episodes in the History of Palaeontology,* 2nd ed. Chicago: University of Chicago Press, 1985.

Ruse, Michael. *The Darwinian Revolution: Science Red in Tooth and Claw,* 2nd ed. Chicago: University of Chicago Press, 1999.

Ruse, Michael, and Joseph Travis, eds. *Evolution: The First Four Billion Years.* Cambridge, MA: Harvard University Press, 2009.

Rutherford, Ernest. *The Collected Papers of Lord Rutherford of Nelson,* vol. 2. New York: Interscience, 1963.

———. *Radioactive Transformations.* New Haven, CT: Yale University Press, 1906.

Rutherford, Ernest, James Chadwick, and Charles Drummond Ellis. *Radiations from Radioactive Substances.* Cambridge: Cambridge University Press, 1930.

Santillana, Giorgio de. *The Crime of Galileo.* Chicago: University of Chicago Press, 1955.

Sapp, Jan. *Genesis: The Evolution of Biology.* Oxford: Oxford University Press, 2003.

Sarton, George. *A History of Science: Ancient Science through the Golden Age of Greece.* Cambridge, MA: Harvard University Press, 1964.

Schrödinger, Erwin. "The Present Situation in Quantum Mechanics." Translated by John D. Trimmer. *Proceedings of the American Philosophical Society,* November 29, 1935, 323–38.

Schwann, Theodor. *Microscopical Researches into the Accordance in the Structure and Growth of Animals and Plants.* Translated by Henry Smith. London: Sydenham Society, 1847.

Seife, Charles. *Alpha and Omega: The Search for the Beginning and End of the Universe*. New York: Penguin, 2004.

———. *Zero: The Biography of a Dangerous Idea*. New York: Viking, 2000.

Selleri, Franco. *Quantum Paradoxes and Physical Reality: Fundamental Theories of Physics*. Dordrecht, Netherlands: Kluwer Academic, 1990.

Shapley, Harlow, ed. *Source Book in Astronomy, 1900–1950*. Cambridge, MA: Harvard University Press, 1960.

Shaviv, Giora. *The Synthesis of the Elements: The Astrophysical Quest for Nucleosynthesis and What It Can Tell Us about the Universe*. Dordrecht, Netherlands: Springer, 2012.

Shaw, H. R. *Craters, Cosmos, and Chronicles: A New Theory of Earth*. Stanford, CA: Stanford University Press, 1995.

Simmons, John G. *Doctors and Discoveries: Lives That Created Today's Medicine*. Boston: Houghton Mifflin, 2002.

Singer, Charles, and C. Rabin. *A Prelude to Modern Science*. Cambridge: Cambridge University Press, 1946.

Smith, Leonard. *Chaos: A Very Short Introduction*. Oxford: Oxford University Press, 2007.

Smith, Robert William. *The Expanding Universe: Astronomy's "Great Debate," 1900–1931*. Cambridge: Cambridge University Press, 1982.

Smocovitis, Vassiliki Betty. *Unifying Biology: The Evolutionary Synthesis and Evolutionary Biology*. Princeton, NJ: Princeton University Press, 1996.

Spadafora, David. *The Idea of Progress in Eighteenth-Century Britain*. New Haven, CT: Yale University Press, 1990.

Stanford, P. Kyle. *Exceeding Our Grasp: Science, History, and the Problem of Unconceived Alternatives*. Oxford: Oxford University Press, 2006.

Stewart, Ian. *In Pursuit of the Unknown: 17 Equations That Changed the World*. New York: Basic Books, 2012.

Stiebing, William H. *Ancient Astronauts, Cosmic Collisions and Other Popular Theories*. Buffalo, NY: Prometheus Books, 1984.

Strickberger, Monroe W. *Evolution*, 3rd ed. Sudbury, MA: Jones & Bartlett, 2000.

Talwar, G. P., and L. M. Srivastava, eds. *Textbook of Biochemistry and Human Biology*, 3rd ed. Delhi: Prentice-Hall of India, 2003.

Taylor, C. C. W. *The Atomists, Leucippus and Democritus: Fragments*. Toronto: University of Toronto Press, 1999.

Thomson, James Oliver. *History of Ancient Geography*. New York: Biblo & Tannen, 1965.

Thorndike, Lynn. *A History of Magic and Experimental Science*, vol. 5. New York: Columbia University Press, 1941.

Tocqueville, Alexis de. *Democracy in America*, vol. 1. London: D. Appleton, 1899.

Topper, David. *How Einstein Created Relativity Out of Physics and Astronomy.* Dordrecht, Netherlands: Springer, 2012.

Trudeau, Richard J. *The Non-Euclidean Revolution.* Boston: Birkhäuser, 1987.

Ts'o, Paul O. P., ed. *Basic Principles in Nucleic Acid Chemistry,* vol. 1. New York: Academic Press, 1974.

Tudge, Colin. *Engineer in the Garden.* New York: Random House, 1993.

Ussher, James. *Annals of the World.* London: E. Tyler, 1658.

Vai, Gian Battista, and W. G. E. Caldwell, eds. *The Origins of Geology in Italy.* Boulder, CO: Geological Society of America, 2006.

Valleriani, Matteo. *Galileo Engineer.* Dordrecht, Netherlands: Springer, 2010.

Van Helden, Albert. *Measuring the Universe: Cosmic Dimensions from Aristarchus to Halley.* Chicago: University of Chicago Press, 1985.

Van Hise, Charles R. "The Problems of Geology." *Journal of Geology* 12, no. 7 (1904): 589–616.

Venter, J. Craig. *Life at the Speed of Light: From the Double Helix to the Dawn of Digital Life.* New York: Viking, 2013.

Verbaal, Wim, Yanick Maes, and Jan Papy, eds., *Latinitas perennis.* Vol. 1, *The Continuity of Latin Literature.* Leiden, Netherlands: Brill, 2007.

Vickers, Brian, ed. *Francis Bacon: The Major Works.* Oxford: Oxford University Press, 2002.

Vitruvius Pollio. *Vitruvius: The Ten Books on Architecture.* Translated by M. H. Morgan. New York: Dover, 1960.

Voegelin, Eric. *History of Political Ideas.* Vol. 6, *Revolution and the New Science.* Columbia: University of Missouri Press, 1998.

Vonk, Jennifer, and Todd K. Shackelford, eds. *The Oxford Handbook of Comparative Evolutionary Psychology.* Oxford: Oxford University Press, 2012.

Wald, Robert M. *General Relativity.* Chicago: University of Chicago Press, 1984.

Wallace, Alfred Russel. *Infinite Tropics: An Alfred Russel Wallace Anthology.* Edited by Andrew Berry. New York: Verso, 2002.

———. *The Wonderful Century: The Age of New Ideas in Science and Invention.* London: Swan Sonnenschein, 1903.

Watson, James D. *The Double Helix: A Personal Account of the Discovery of the Structure of DNA.* New York: Scribner, 1993.

Webster, Charles, ed. *The Intellectual Revolution of the Seventeenth Century.* New York: Routledge, 2011.

Wegener, Alfred. "The Origin of Continents and Oceans." *Living Age,* 8th series, vol. 26 (April/May/June 1922): 657–61.

———. *The Origin of Continents and Oceans.* Translated by John Biram. New York: Dover, 1966.

Weinberg, Steven. *Dreams of a Final Theory: The Scientist's Search for the Ultimate Laws of Nature.* New York: Vintage, 1994.

————. *The First Three Minutes: A Modern View of the Origin of the Universe*, 2nd ed. New York: Basic Books, 1993.

Whiston, William. *A New Theory of the Earth, from Its Original, to the Consummation of All Things*, 5th ed. London: John Whiston, 1737.

Williams, Malcolm. *Science and Social Science: An Introduction*. London: Taylor & Francis, 2002.

Wilson, Edward O. *Letters to a Young Scientist*. New York: Liveright, 2013.

————. *On Human Nature*. Cambridge, MA: Harvard University Press, 2004.

————. *The Social Conquest of Earth*. New York: W. W. Norton, 2012.

————. *Sociobiology: The New Synthesis*. Cambridge, MA: Harvard University Press, 1975.

Wolpert, Lewis. *The Unnatural Nature of Science*. Cambridge, MA: Harvard University Press, 1992.

Index

Page numbers in *italics* refer to illustrations and figures.

geometry, 21–29, 43, 70, 218–19, 223
George III, king of England, 166
Gerard of Cremona, 43
Germany, Germans, 45, 83, 85, 112, 120, 144, 157–58, 177, 179, 180, 186, 189, 190, 206, 226, 228, 232
Gibraltar, Strait of, 58
Gilluly, James, 150
glaciers, 142, 149
Gleick, James, xviii, 213, 252, 255–56
glossopetrae, 107
God:
    as creator, 120, 129, 132, 134–35, 161, 167, 179–80, 199, 216–17
    and divine revelation, 76
    in Great Chain of Being, 18
    intervening in natural order, 112, 217
    in natural philosophy, 100
Goddard, Dr., 90
gods, Greek, 3–7, 9, 12–13, 18, 23, 32
gold, 85
Gorgias of Heraclea, 6
Göttingen, University of, 219–20
Gould, Stephen Jay, xviii, 155, 199, 206, 208–11, 252, 255
grasshoppers, 178, 189
gravitas (heaviness), 97
gravity:
    in four-dimensional universe, 219–20
    in general theory of relativity, 223
    Newton's theory of, 97–102, 215–17
Great Chain of Being, 18
Great Deluge, 124–25
Great Instauration (Bacon), 57–60
Greece, ancient, 32, 240
    atomic theory in, 226
    change in, 16–20, 118
    chemistry in, 84
    European rediscovery of scientific legacy of, 43–44
    geography in, 105
    language and vocabulary of, 14–15, 19, 47

mathematical measurement of universe in, 21–31, 38–39
    number system of, 30–31
    origins of universe in, 3–8, 38, 43, 106, 118
    philosophy of, 10–15, 16–20, 25–27, 32–34
Greek Science: Its Meaning for Us (Farrington), xvii
Greenland, Wegener's expeditions to, 144, 146
Gribbin, John, 248
Guericke, Otto von, 83
Guicciardini, Nicolò, 98n

Haeckel, Ernst, 163, 177
Haldane, J. B. S., 200, 201
Hamilton, William D., 202, 205
Hamilton's Rule, 202–3
Hammond, J. C., 145
haplodiploidy, 202
Harmony of the Spheres, 23
Harvard University, 154, 206
Harvey, William, xix, 53, 63–69, 70, 92
    reading sources for, 69
Hawking, Stephen, 52, 248
hawkweed, 176
healing, 5–6, 68
heart, nerves and, 72–73
heart function, 65–68
Heath, Thomas, 31
"heavy matter," 37, 46, 71
Heisenberg, Werner, 232–33
Heisenberg Uncertainty Principle, 233
heliacal rising, 4
heliocentric model, 29–30, 46–52, *48*, 55, 76–78, 96–98, 216; *see also* solar system
hemipterans, 189
hemophilia, 190–91
Henking, Hermann, 189
Henslow, John, 167
Heraclitus, 11
heredity, *see* inheritance
heresy, 76
Herodotus, 32